TECHNOLOGICAL HAZARDS

LANCASTER
UNIVERSITY

LIBRARY

Donald J. Zeigler
Department of Political Science
and Geography
Old Dominion University
Norfolk, Virginia

James H. Johnson, Jr.
Department of Geography
University of California
Los Angeles

Stanley D. Brunn
Department of Geography
University of Kentucky
Lexington

RESOURCE PUBLICATIONS
IN GEOGRAPHY

Library of Congress Card Number 83-22356
ISBN 0-89291-173-5

Library of Congress Cataloging in Publication Data

Zeigler, Donald J., 1951-
 Technological hazards.

 (Resource publications in geography)
 Bibliography: p.
 1. Technology — Social aspects. 2. Disasters.
3. Risk. I. Johnson, James H., 1954-   II. Brunn,
Stanley D.   III. Title.   IV. Series.
T14.5.Z45 1983   363.1   83-22356
ISBN 0-89291-173-5

Publication Supported by the A.A.G.

Graphic Design by CGK; cover art by N. L. Diaz, University of California at Los Angeles.

Printed by Commercial Printing Inc.
State College, Pennsylvania

Foreword

For the technological optimist, technology will not only enable us to mitigate the perils of natural hazards, but will also enable us to correct the unforeseen problems created by science and technology itself. In this book, Don Zeigler, Jim Johnson, and Stan Brunn argue that technology has created a new category of hazards — parallel in some ways to the natural hazards that have received much attention by geographers, but different as well. For the technological pessimist, these new hazards are compounding — solving one problem is more than likely to create many others.

Contemporary geographical views of natural hazards more clearly address the potential synergism between the social processes and patterns on the one hand, and extreme natural events on the other, that create 'natural hazards.' Technology viewed as hazard will enhance this viewpoint: here human society not only creates the fundamental pattern of vulnerability but also initiates the hazardous event itself.

Because technological hazards have important spatial as well as social and temporal dimensions, it is appropriate that geographers have contributed significantly to thought on this topic. The *Resource Publications in Geography,* sponsored by the Association of American Geographers, present contemporary views of geographers on issues of public as well as professional importance. The Association's purpose is to advance studies in geography and to encourage the application of geographic research in education, government, and business. The *Resource Publications* traces its origins to the AAG's 1968-1974 *Resource Papers* and continues an original commitment to bringing modern geographical thought to students and to issues of public policy and current social relevance.

Although this book was developed by the AAG, the ideas are the authors' and do not imply AAG endorsement. The editor and advisory board urge careful study of *Technological Hazards* — for the sake of mankind as well as to reaffirm our commitment to use our geographical expertise to create a better world.

C. Gregory Knight, *The Pennsylvania State University*
Editor, Resource Publications in Geography

Resource Publications Advisory Board

George W. Carey, *Rutgers University*
James S. Gardner, *University of Waterloo*
Charles M. Good, Jr., *Virginia Polytechnic Institute and State University*
Risa I. Palm, *University of Colorado*
Thomas J. Wilbanks, *Oak Ridge National Laboratory*

Preface

Technological hazards have assumed a place alongside natural hazards as major social problems and as significant foci of problem-oriented geographic research. In writing this monograph it has been our objective to suggest a unifying geographic structure for the study of technological hazards. We have tried to draw upon a wide variety of specific hazards in illustrating the models and concepts we develop. It has not been our intention to offer an encyclopedia of technology's negative impacts, but it has been our desire to cite a broad array of the recently published literature in the hazards realm in anticipation that this monograph might be used as a springboard for additional reading and investigation. We believe there is great potential for a multitude of contributions to society for those pursuing research in the technological hazards arena.

All three of us have been involved in special hazard investigations, particularly with respect to nuclear power generation and the problems which it presents. Examples include the locational conflict over radioactive waste disposal and the public response to emergencies at nuclear generating stations. In Chapter 5 we take the opportunity to discuss some of our research on evacuation, a process we first studied in the aftermath of the accident at Three Mile Island. In addition to our research, we have also taught graduate and undergraduate hazards seminars at four different universities, and we make it a practice to cover technological hazards in our introductory classes. We hope, therefore, that this monograph might serve both the growing number of specialized courses in hazards and the general introductory classes with a problem-oriented focus. We also see it as having applicability to courses in social geography, political geography, and environmental problems.

Many of the models, concepts, and case studies we discuss are presented graphically. In fact, we like to think that the monograph revolves around the many maps and diagrams we have included. For preparing these graphics we have many people to thank. At the University of Kentucky, we extend our appreciation to Gyula Pauer, Sr., director of the Cartography Laboratory, and his assistants Joe Watts, Cinda Taylor, Joe Carter, Sherri Snell, and Gyula Pauer, Jr. At the University of California, Los Angeles, we extend our appreciation to Noel Diaz, who is responsible not only for many of the graphics but also for this Resource Publication's cover design, and to Suzanne LeBlanc. We would also like to acknowledge the UCLA Academic Senate for providing funds for some of the maps. In addition, both Donald K. Emminger and Peter M. Rhodes contributed their graphic skills to our efforts. For reviewing parts of the manuscript and providing other helpful input, we must thank Dr. Russell B. Capelle, a geographer with the American Trucking Associations, and Dr. Justin C. Friberg, a colleague at Old Dominion University. For typing the manuscript several times over, we thank Mrs. Evangeline Buggs and Ms. Linda Stephenson. Finally, for the funding provided to carry out the Long Island evacuation survey discussed in Chapter 5 we must extend our appreciation to Peter F. Cohalan, Chief Executive Officer of Suffolk County, New York.

Donald J. Zeigler
James H. Johnson, Jr.
Stanley D. Brunn

Contents

List of Figures

List of Tables

1

Love Canal, Three Mile Island, and Nuclear Freeze

On August 7, 1978, the President of the United States declared a national emergency at Love Canal in Niagara Falls, New York, where 43 million pounds of toxic chemicals had been dumped between 1947 and 1952 (Epstein *et al.* 1982). This action marked the first time that a technological rather than a natural hazard had been directly responsible for such a declaration. The on-going Love Canal incident established the problem of toxic chemical disposal, neglected for the past quarter of a century, as one of the leading environmental issues of the 1980s (Brown 1981; Levine 1982). Since 1976, when chemicals buried at Love Canal began to surface after several years of heavy rain, the American people have become aware that 4½ million chemical compounds have been registered with the American Chemical Society. Only a fraction of the chemicals being manufactured have been tested for carcinogenic qualities; toxic waste dumps dot the landscape of almost every heavily populated area in the United States; and the ground water resources of these areas and the people who inhabit them are seriously threatened.

One year later, as the residents of Love Canal were agitating for recognition of the seriousness of their problem and an expansion of the zone of evacuation, a technological disaster brought south central Pennsylvania to the attention of the world. On the morning of March 28, 1979, one of Metropolitan Edison's two nuclear generating units at Three Mile Island experienced a chain of mechanical failures and human errors which resulted in the release of several puffs of radioactive steam and the potential for an even larger disaster (Martin 1980; President's Commission 1979). The uncertainty of the moment resulted in the voluntary evacuation of more than 144,000 inhabitants of the immediate area (Flynn 1979) and the commencement of a newly revived debate over the use of nuclear technology for generating electricity, a debate which Ralph Nader has suggested may become a "technological Vietnam."

On the national front, as the second Strategic Arms Limitation Treaty died during the waning years of the 1970s, debates over neutron bombs that would kill people while inflicting minimal damage on property (Westing 1978) gave way to debates over a mutual United States-Soviet Union freeze in nuclear weapons development and over the deployment of new U.S. medium-range missiles in Europe. Meanwhile, the world's super powers, with more than 17,000 nuclear warheads between them, prepare for the total annihilation which they hope will not come. Smaller nations, often characterized by unstable political regimes, developing economies hungry for investment capital, and intolerable neighbors on their frontiers, are adopting nuclear technologies for both energy production and "defense."

These three issues — hazardous waste disposal, the development of nuclear power, and preparations for nuclear war — are three of the many technological hazards which demand our attention in the 1980s. They represent technologies whose potential negative impacts are severe, long-lasting, and potentially global in scale. They are but a few of the total number of technological hazards which society must learn to cope with between now and the twenty-first century. Technological hazards have risen in number and diversity, in magnitude and scope of potential harm, and in the public's con- sciousness to the extent that the risks of technological development and applications are among the most hotly debated of contemporary issues. While technological hazards are as old as technology itself, the cumulative effects of technological advance have spawned the development of a society where large-scale technology and the decisions regarding its deployment are made not by individuals but by an elite who operate under the neo-Mackinderism, "who controls technology controls the world."

Society's Interest in Technological Hazards

The view of technology as the cornerstone of all progress, the "technologic ethic" (Tichener *et al.* 1971), has prevailed in the past in large part because of "the pressure of competition, rapid population growth, and resource depletion" on the one hand and "a vision of an even better quality of life, of indefinite progress, novelty, and inventive discovery" on the other (Carpenter 1978:97). Since the 1950s, however, opinion polls have shown that the American public is becoming more concerned about the threat of unbridled technological growth as it may affect public health, the environment, job security, individual privacy, the participation in democratic government, and the overall quality of human life (Withey 1959; Taviss 1972; Bunge 1973; Goldman *et al.* 1973; Mazur 1973; Opinion Research 1974; Etzioni and Nunn 1974; Cotgrove 1975; LaPorte and Metlay 1975a, 1975b, 1976; Norman 1981). Moreover, some members of the scientific community (Nelkin 1975; Singer 1977; King 1977) have expressed similar concerns about the physical, ethical, and economic consequences of continued research and development of what Graham (1978) describes as destructive (*e.g.* military weapons), slippery slope (*e.g.,* genetic engineering), and economically ex- ploitative (*e.g.,* supersonic transport) technologies.

We have come to realize that technology is essentially a double-edged sword (Weisskopf 1980). Some see it as the source of humankind's strength (Wiesner 1973) and worry that overcautious attitudes toward new technologies may paralyze scientific endeavor, thus increasing the risks which we face in life (Wildavsky 1979). Others see technology as generating hazards which may adversely affect the earth's social, economic, and environmental subsystems. They believe that we too often look to technology to solve human and environmental problems, not realizing that they may displace human beings, dehumanize the individual and workplace, and upset fragile environmental balances.

In a very controversial treatise (Deveault 1981; Gordon 1981), Rifkin goes so far as to suggest that the second law of thermodynamics, the entropy law, "destroys the notion that science and technology create a more ordered world" and maintains that each new technology "creates a temporary island of order at the expense of greater disorder to the surroundings" (1981:6, 81). Toffler (1980:151) refers to those who question the value of technology as "techno-rebels" and sees their role as agents of the "Third Wave." These social visionaries argue for discerning, selective development of technologies responsive to — rather than directing — long-range social and ecological goals. Schumacher (1973:153-154) foresees the technology of the future as "a technology to which everyone can gain admittance and which is not reserved to those already rich and powerful." He calls it *intermediate technology.* It would be "conducive to decentralization, compatible with the laws of ecology, gentle in its use of scarce resources, and designed to serve the human person instead of making him the servant of machines." It is "technology with a human face," — called *appropriate technology* by many.

During the past two decades society has seen a multitude of responses to technological threats. Organizations have been established, books written, and legislation passed to control the uses and misuses of technology or alleviate its impact on the environment (Tables 1-3). In 1972, the U.S. Congress established the Office of Technology Assessment because it felt it was essential that "the consequences of technological applications be anticipated, understood and considered in determination of public policy on existing and emerging national problems" (P.L. 92-484). Over the same time span, technology assessments have grown worldwide (Gibbons 1983).

Man-made hazards have also been covered extensively in the media. Hardly a week passes without newspaper coverage of another oil spill, train derailment, chemical explosion, or discovery of another hazardous waste dump. In his book *Technology and Social Shock,* Lawless (1977) compiled a list of 100 hazards of technological origin which were important enough to make the headlines of the national press between 1945 and 1972. Included are case studies of nerve gas and radioactive waste disposal controversies, mercury and DDT pollution, foaming and enzyme detergents, the Thalidomide tragedy, the cyclamate affair, and synthetic turf and football injuries. From U.S. and foreign periodicals between 1970 and 1978, Gladwin (1980) identified 3000 environmental conflicts over the siting of industrial facilities in forty different countries.

In an effort to measure disaster magnitude, Canadian geographer Harold Foster (1976) devised a logarithmic scale based on a stress index which he calculated for a variety of calamities. In his category of "major catastrophes" were the Black Death (bubonic plague) of fourteenth century Europe and two deliberate, anthropogenic armed conflicts, World Wars I and II. Technological advances in the medical sciences have almost eliminated the likelihood of another Black Death, but advances in military technology have far from eliminated the possibility of World War III. Instead, cumbersome geopolitical structures seem pushed to the breaking point in trying to keep thermonuclear and other weapons under control. In fact, should anything like the medieval plagues sweep across the ecumene of the future, they are likely to be the result of germ warfare and not natural causes. Other exemplary technological hazards which were ranked by Foster as "catastrophes" and "disasters" included the 1971 mass poisoning from fungicide-treated grain in Iraq, the dropping of the atomic bomb on Hiroshima, the sinking of the Titanic, a train derailment in a tunnel, and a munitions ship explosion in Nova Scotia. To this list we might add a disaster about which little is known, the 1957-1958 explosion of high-level radioactive wastes at a military plutonium production site near Chelyabinsk in the Soviet Union. Recent maps have deleted the names of about 30 towns and shown modifications of stream patterns in the area (Medvedev 1979; Trabalka *et al.* 1980). Among these acute technological hazards we should also include the chronic hazards of pollution and occupational disease which result in physical and social consequences distributed over a long period of time and more difficult to trace to their original causes.

TABLE 1   ORGANIZATIONAL RESPONSES TO TECHNOLOGICAL HAZARDS

Organization Name Headquarters Year Founded	Purpose
Educational Foundation for 　Nuclear Science Chicago, Illinois 1945	Promotes study of impact of science and technology on public affairs; publishes *Bulletin of the Atomic Scientists.*
Campaign for Nuclear Disarmament London, England 1958	Supports nuclear disarmament and non-military solutions to conflicts between nations.
World Future Society Bethesda, Maryland 1966	To contribute to a reasoned awareness of the future and the importance of its study.
Science for the People Cambridge, Massachusetts 1969	To create a society based on human needs rather than profit, and to create a science that serves all the people.
Union of Concerned Scientists Cambridge, Massachusetts 1969	Advocate organization concerned about the impact of advanced technology on society.
Friends of the Earth San Franciso, California 1969	Works to generate among people a new responsibility to the environment
Save the Whales Washington, D.C. 1971	To publicize the dilemma of the great whales, to save them from extinction, and to work for whale regeneration.
Alternative Energy Resources 　Organization Billings, Montana 1974	To encourage energy efficiency and the transition from conventional to renewable energy resources.
Citizens Energy Project Washington, D.C. 1973	Research concerning renewable energy technologies, anti-nuclear issues, and community-appropriate technology.
Worldwatch Institute Washington, D.C. 1974	Encourages a reflective, deliberate, anticipatory approach to global problem-solving.
Intermediate Technology Menlo Park, California 1975	Seeks to develop small-scale, low-cost technologies more compatible with the environment and human needs.

TABLE 1   ORGANIZATIONAL RESPONSES TO TECHNOLOGICAL HAZARDS

Organization Name Headquarters Year Founded	Purpose
National Center for Appropriate Technology Butte, Montana 1976	To develop and apply technologies appropriate to the energy-related needs of low-income people and communities.
International Human Powered Vehicle Association Claremont, California 1976	To further the development of land, water, and air varieties of human powered vehicles.
Appropriate Technology International Washington, D.C. 1977	To assist developing countries in implementing technologies appropriate to their economic and social circumstances.
International Project for Soft Energy Paths San Franciso, California 1978	Dedicated to publc awareness of soft energy — alternate forms of energy that are efficient, renewable, and environ- mentally benign.
Karen Silkwood Fund Washington, D.C. 1978	Seeks to protect worker and public from willful negligence on the part of the nuclear industry.
Waste Watch Washington, D.C. 1979	Concerned with national resource waste problems; promotes citizen participa- tion in waste issues.
Children's Campaign for Nuclear Disarmament Plainfield, Vermont 1981	Letter-writing campaign by children who oppose the nuclear arms race.
Global Futures Network Toronto, Ontario 1981	Promotes the growth of global under- standing and co-operative efforts toward a new stage of planetary evolution.
High Technology Professionals for Peace Cambridge, Massachusetts 1981	To educate members and non-members about the nuclear arms race and the capabilities of nuclear weapons.
International Green Alliance Santa Ana Heights, California 1981	To create a world in which the quality of life is more important than a standard of living or nuclear superiority.

TABLE 1   ORGANIZATIONAL RESPONSES TO TECHNOLOGICAL HAZARDS

Organization Name Headquarters Year Founded	Purpose
Nuclear Weapons Freeze Campaign St. Louis, Missouri 1981	Seeks a bilateral U.S.-U.S.S.R. agreement to stop production, testing, and development of nuclear weapons.
Society for Risk Analysis Bethesda, Maryland 1981	To study and understand on a scientific basis the risks posed by technological development.
Great Lakes United Watertown, New York 1982	To promote the conservation and enhancement of the Great Lakes ecosystem.
National Resource Recovery Association Washington, D.C. 1982	To promote the development of resource recovery facilities, district heating, and urban waste energy systems.
Social Scientists Against Nuclear War New York, New York 1982	Seeks to end the threat of nuclear war and construct an alternative program for national defense.
People's Medical Society Emmaus, Pennsylvania 1982.	Promotes citizen involvement in the American health care system.

Source: Akey 1982.

Is the number of technological hazards increasing? Kates (1977b:5-7) suggests that our increasingly sophisticated ability to detect and identify hazards has led us to believe that the number of technological threats is increasing just as fast. His survey of the *New York Times* and *Environment* magazine, however, showed that between 40 and 50 different hazards are discovered annually, a range which has been consistent during recent years. Gladwin (1980; 250) refers to the early 1970s as the "age of alarmed discovery and euphoric enthusiasm" and notes that the reported conflict over industrial facility siting has remained about constant since 1972. Regardless of their rate of increase, however, technological hazards already impose monumental costs on society. The joint research project carried out by the Clark University Center for Technology, Environment, and Development, along with Decision Research (1979) estimated that between 15 and 25 percent of all deaths in the United States are the result of technological hazards. The price we pay to cope with the hazards of technology ranges between $200 and $300 billion per year, and at least half of the increased rate of species extinction may be explained by technological factors. The methodologies employed in generating these findings are discussed by Harris *et al.* (1978).

An analogy with a well-known model of human motivation may help us to understand society's continuing interest in hazards of both natural and technological origin.

TABLE 2 LITERARY RESPONSES TO TECHNOLOGICAL HAZARDS

Book	Author	Year
The Failure of Technology	F. Juenger	1956
Silent Spring	R. Carson	1962
The Technological Society	J. Ellul	1964
Future Shock	A. Toffler	1970
The Closing Circle	B. Commoner	1971
Small is Beautiful	E. Schumacher	1973
The Future of Technological Civilization	V. Ferkiss	1974
Autonomous Technology	L. Winner	1977
The Technological Conscience	M. Stanley	1978
The Technology Trap	L. Moser	1979
The Social Control of Technology	D. Collingridge	1980
The Technological System	J. Ellul	1980
The Third Wave	A. Toffler	1980
Entropy	J. Rifkin	1981
The God That Limps	C. Norman	1981
Catastrophe or Cornucopia?	S. Cotgrove	1982
The Deindustrialization of America	B. Bluestone and B. Harrison	1982
What's Wrong With Our Technological Society — And How to Fix It	S. Ramo	1983
Despair and Nuclear Power in the Nuclear Age	J. Macy	1983

Abraham Maslow, the noted psychologist, defined a hierarchy of human needs (Figure 1) which established the fulfillment of physiological and safety needs as necessary preconditions for advancement up the hierarchy to self-actualization, the tendency for a person to aspire "to become actualized in what he is potentially" (Maslow 1970:35-36). Each order of individual needs has a counterpart at the community level. The safety needs of the individual may be translated into the need for a community environment free from hazards and civil disturbances. Traditionally, the safety needs have been met by the formation of police and fire departments, and more recently by emergency preparedness agencies, hazard-conscious community planning, and risk management institutions. Hazards demand our attention because they threaten the foundation on which community well-being is and will be based. Technological hazards in particular may directly compromise the safety, health, and property of individuals and may indirectly threaten their well-being by inducing stress and degrading the ecosystems which provide a community with the physiological necessities of life: food, air, and water. If community-actualization is the ultimate goal of local government, lower-order community needs must first be met. Otherwise, they will continue to preoccupy the attention of governing boards and prevent them from moving on to fulfill higher order community needs. Unfortunately, as the scale of technological impact increases, community hazards are more likely to originate outside the local area and beyond the control of local decision-making authorities.

TABLE 3   LEGISLATIVE RESPONSES TO TECHNOLOGICAL HAZARDS

Federal Act	Year
Solid Waste Disposal Act	1967
National Environmental Policy Act	1969
Occupational Safety and Health Act	1970
Clean Air Act Amendments	1970
Federal Water Pollution Control Act	1972
Ocean Dumping Act	1972
Noise Control Act	1972
Consumer Product Safety Act	1972
Federal Insecticide, Fungicide, and Rodenticide Act	1972
Technology Assessment Act	1972
Endangered Species Act	1973
Safe Drinking Water Act	1974
Hazardous Materials Transportation Act	1974
Resource Conservation and Recovery Act	1976
Toxic Substances Control Act	1976
Surface Mining Control and Reclamation Act	1977
National Energy Policy Act	1978
Hazardous Liquid Pipeline Safety Act	1979
Environmental Response, Compensation, and Liability Act	1980
Nuclear Waste Policy Act	1982

Contrasting Natural and Technological Hazards

Most of the hazard research conducted in geography and related fields has been specific to natural hazards (Mitchell 1974b; White 1974; White and Haas 1975; Burton *et al.* 1978; Quaranteli 1978; and Mitchell 1979). Burton and Kates (1964:413) have defined natural hazards as "those elements in the physical environment, harmful to man and caused by forces extraneous to him." They also formulated the most commonly used hazard typology, a genetic classification which divides natural hazards into those of geophysical and those of biological origin. The school of hazard research fathered by Gilbert White has concentrated on the geophysical hazards, while medical geographers have developed other methodologies for studying such biological threats as malaria, schistosomiasis, and cancer. The growing interest in geophysical hazards has witnessed studies of floods (White 1945; Kates 1962; Mileti and Beck 1975; Erikson 1976), droughts (Saarinen 1966; Heathcote 1969; Brooks 1971), severe storms (Sims and Saarinen 1969; Sims and Bauman 1972; Baker 1979), volcanic eruptions (Rees 1979), and erosion (Mitchell 1974a). During the 1970s technological threats and disasters have commanded a greater share of attention as research topics for geographers, sociologists, and psychologists (Mason 1971, 1972; Kates 1977b; Quaranteli *et al.* 1978; Olson 1979; Slovic *et al.* 1979). Probably the most important centers in the United States for technological hazard research have been the Center for Technology, Environment and Development at Clark University, the Disaster Research Center at Ohio State University, and the Decision Research Corporation in Eugene, Oregon. As

FIGURE 1 Hierarchies of Individual and Community Needs (Individual needs based on Maslow 1970)

interest in the hazards of technology has grown, however, studies have been conducted by social scientists throughout the nation.

A comprehensive hazard theory needs to distinguish between natural and technological hazards. Public policies which seek to manage and mitigate hazards must take into consideration the fact that people perceive many technological hazards differently from natural hazards. If perceptions differ, behaviors are likely to differ as well. The following factors may help us understand the basis for these differing perceptions and behaviors.

First, natural hazards have been with humankind throughout our evolutionary history. The range of natural hazards has been known and the periodic excesses of nature have been recorded. Technological disasters have also been around since paleolithic beings began using stone weapons. However, the technological hazards of the techno-primitive past are dramatically different from the technological hazards that characterize today's society. Paleolithic beings had experience with floods and volcanoes but not with toxic chemicals and nuclear energy. An historical record may be drawn upon to help predict natural hazards but no historical record has been amassed for many newly developed technologies. Since there is usually a period of smooth performance during the initial stages of a technology's implementation, it may be unwise to base predictions of the future on an "historical record" that extends back only a few years or even a few decades.

Second, the emphasis in natural management is on prediction of the hazard rather than its elimination. In the technological realm, there is an unacknowledged faith that any hazard can be eliminated from the system. We expect natural hazards, but because "technology is not built to break down" we never really expect technological ones (Baum *et al.* 1983:342). We therefore seldom eliminate threatening technologies, preferring instead to apply more technology to make the system fail-safe. Harmful food additivies and drugs, flaws in weapons systems, and imperfections in the design of nuclear power plants are countered by programs to work with the product or within the system to make it safe and publicly acceptable. Permanently removing dangerous products or dismantling harmful technological systems has seldom occurred. However, "the new social values that place a premium on safety and environmental preservation" (Norman 1981:30) may increasingly lead to the rejection of technologies that demean human beings and debilitate their environments. In 1977, for instance, Dow Chemical abandoned, because of community outcry, a petrochemical complex that would have employed 1000 workers in the San Francisco area (Gladwin 1980: 246). Likewise, in Austria a completed nuclear power plant sits idle because the voters in a national referendum cast their vote, albeit by the slimmest of margins, against its operation (Patterson 1979).

Third, natural hazards differ from technological hazards in scope of potential harm, both spatially and temporally. Technological hazards of the highest order, ranging from the seemingly innocuous use of chlorofluoromethanes (aerosol propellants) to all-out nuclear war, have a potentially global scale of impact. Natural hazards, on the other hand, may be regional in scope but seldom threaten the entire world. The eruption of Tambora in Indonesia, which spewed so much volcanic ash ino the atmosphere that 1815 was known as the "year without a summer," seems to have been one of the very few natural disasters with major global consequences. Tambora's impact, however, was relatively short-lived. Some technological hazards, on the other hand, may induce long-lasting or permanent changes in the earth's environment. A recent report by the U.S. Environmental Protection Agency (1983) warned that the earth's temperature could increase 3.6°F by the year 2040 as a result of a buildup of atmospheric carbon dioxide, a by-product of fossil fuel combustion. The report further suggested, arguably, that the warming trend is not only inevitable but irreversible. At this point in time even a total ban on fossil fuel combustion would only delay the warming effect a few years. Unlike natural hazards which may endure for a decade or so at most (e.g., regional droughts), carbon dioxide-induced temperature changes may be permanent. As Baum *et al.* (1983) observed, many technological hazards are not over-and-done-with as are natural hazards; there is no well-defined "low point," a time at which the worst is over and life returns to normal. In the case of radioactivity and toxic chemicals the feeling remains that the worst may be yet to come. In severe accidents, in fact, the effects of radiation or toxic chemicals may not subside for hundreds or even thousands of years.

Fourth, natural disaster agents are almost always visible while some technological hazards agents are invisible, yet lethal. The different behavioral responses to floods and nuclear power plant accidents may, in part, be related to an individual's personal ability to size up the hazard through direct observations of the disaster agent and its impact. Asbestos dust and many toxic chemicals look quite harmless, but when inhaled or ingested serious somatic effects may result. Likewise low doses of radioactivity are imperceptible, yet the threat they pose is severe and long lasting. Later effects of ionizing radiation include lens opacities and cataracts, infertility, birth defects, stymied growth and development, chromosomal aberrations, and cancer, especially leukemia (Beebe 1982:36-38).

Despite their differences, natural and technological hazards are often dynamically interlinked; natural hazards may cause, compound, or intensify technological ones. Several California nuclear reactors are located near active fault lines; offshore oil rigs in the Gulf of Mexico are in the path of tropical storms; and heavy rains may intensify the hazards associated with the accidental release of radioactive materials from a nuclear power plant. On the other hand, the hazards of technology may be distributed over large areas by the earth's natural systems. For instance, the sulfur oxides produced as a by-product of fossil fuel combustion are felt as acid rain in New England, eastern Canada and Scandinavia because of circulation patterns in the upper atmosphere. Similarly improperly disposed toxic chemicals become regional threats because they are prone to disperse into groundwater aquifers. In a parallel vein, overdependence on technology to eliminate the hazards of extreme natural events may entice people into settling areas that would be best devoted to less intensive uses. In 1983 floods on the Colorado River quickly filled up several reservoirs, requiring the release of water before they overflowed. Downstream, floodplains were inundated along with their residents who felt safe in building there because technology had brought the Colorado "under control." Hazard planning and remediation at local, state, and national levels needs to address the interactions of natural and technological systems whenever appropriate.

Technological Hazard Zones: Geometry and Geography

Hazards may originate in the natural or technological environment, or they may arise from a lifestyle which is actively or passively dangerous. Lifestyle hazards may be associated with backyard swimming pools in which more than 350 people drowned in 1982, with the high-fat "affluent diet" which is so typical of the developed world (Eckholm 1977), or with behaviors such as hitch-hiking and burglary, to name just a few. Technological hazards, on the other hand, originate in the technological environment, the shell which humankind has erected to protect itself from the forces of nature and make it easier to fulfill wants and needs. Technological hazards complement natural and lifestyle hazards in defining the "hazardousness" of a place or region. "The earth's surface is an intricate risk mosaic," noted Foster (1980:43). The following typology of technological hazards, and the hazard zones which they create, makes it possible to develop that analogy further in Chapter 4.

As portrayed in Figure 2, the hazards of technology may be divided into public hazards and private hazards; the former present a threat to the public at large and the latter to individuals alone. Since most technologies are employed to more efficiently achieve economic goals, the scheme presented for public technological hazards is based on the model of an economic system. Some hazards originate with the production of primary commodities or manufactured goods; others in the transportation and transmission sector; and still others arise as a result of the consumption of products which the economic system makes available.

On the other side of the diagram, private hazards are those which are assumed by individuals, sometimes unknowingly, as a result of their occupation or their personal susceptibilities to the negative side-effects of technologies. Occupational hazards are hazards of the workplace; they may evolve from the job itself or from the workplace environment. Textile workers do not suffer from cancer and byssinosis ("brown lung disease") because of the jobs they perform but because the environments in which they work are charged with acrylonitrile (a chemical used to manufacture synthetic fibers) in the first instance and cotton dust in the second (Behr 1978). In general, the risks associated with occupational hazards have been underestimated because the emphasis has been on occupational safety rather than occupational disease. Accidental injury has been dealt with in the 1970 Occupational Safety and Health Act which requires employers to provide a workplace "free from recognized hazards that are causing or are likely to cause death or serious physical harm to employees" (Berman 1978:33). Hazards which result in occupational disease, however, are largely invisible

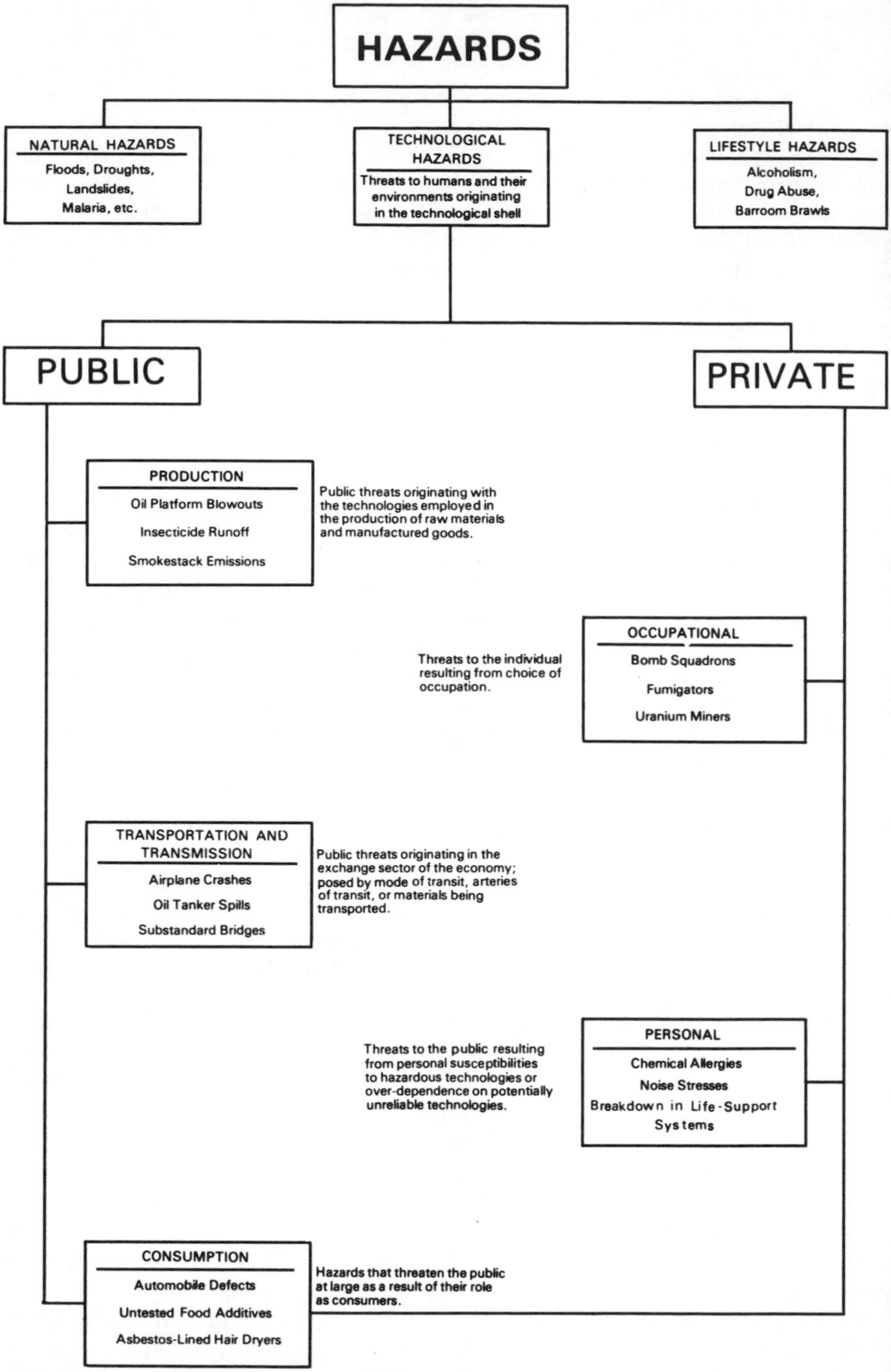

FIGURE 2   A Typology of Technological Hazards

over the short term; employee compensation for diseases which are the result of long-term low level exposure to harmful substances is more difficult to secure. Yet, a federal study has estimated that 20 to 40 percent of cancer deaths can be linked to on-the-job exposure (Berman 1978:46-51). Interviews conducted by Nelkin (1983) with workers who are routinely exposed to toxic chemicals in their jobs revealed a high level of anxiety and a feeling of powerlessness to exercise any control over the hazard.

Associated with each of the public technological hazards are hazard zones, regions which are defined according to the level of risk to which the public is subjected given the proximity of particular technologies. The hazard may result from a malfunction in the technological system or from a negative external effect of a technology which is the by-product of its application. Each of the three types of public hazards generates a different hazard zone geometry. The hazard zones of production hazards tend to be area-wide with the highest risks experienced by those closest to the point of production. The hazard zones of transportation and transmission hazards tend to be linear in shape with some parts of the route (such as intersections and terminals) associated with higher risks than others (Miller and Kaufman 1978; Marino and Becker 1978). Most oil spills from Great Lakes tankers, for instance, occur in harbors and their approaches (Keillor 1980). The hazard zones of consumption hazards tend to be punctiform in nature since risk is not a function of geographical location but instead of consumer behavior. The geographic pattern which evolves mirrors the distribution of consumers in space. Examples of these hazard zone geometries are provided in Figure 3. To this typology must be added a volumetric dimension since many of the impacts of technology are three dimensional, extending from the depths of the ocean to edge of the atmosphere.

The hazard zone itself should not be confused with the impact zone. The former is the area at risk and the latter the area where the risk has been realized; one is potential and the other actual. Determining where a disaster is likely to occur is a geographical problem of considerable importance. The impact zone of nuclear power plants, for instance, would be determined by the site of the reactor and by the magnitude of an accident. Transportation and transmission hazards on the other hand will occur somewhere along the routes which make up the network, but exactly where becomes a matter for probabilistic conjecture since the technology is mobile rather than fixed in space. The impact zone for a particular technology is most often a subset of the hazard zone. This presents a problem for emergency planning in some instances.

Several million packages of radioactive materials traverse the nation's highways yearly; the volume is growing (Church and Norton 1981). The transportation of these wastes is done with stringent safeguards. Nevertheless, the chance of an accident occuring somewhere along the linear hazard zone is always present. Only a handful of sites around the United States will accept radioactive wastes; this means that the distances travelled may be extremely long. Low level wastes from Three Mile Island, for instance, have been transported across the continent to storage facilities in Richland, Washington, and Idaho Falls, Idaho, to the consternation of local government officials along the way who believe that such shipments present a threat to their communities. Planning for accidents which could occur anywhere along a 3000-mile hazard zone presents many problems which are different from accident planning for site-bound nuclear power plants.

Once a technological hazard agent has been identified, management strategies must be based on an enumeration, ranking, and quantification of its characteristics and

HAZARD ZONE
GEOMETRIES

The hazard zone is the area
at risk given:

(1)   A malfunction in the
       technological system; or
(2)   Deleterious by-products
       of the application of
       a particular technology.

PRODUCTION HAZARDS

Example:
Radiation Leakage at a Nuclear Power Plant

Geometry:
Areal

Contoured Risk Surface:
Each point within the hazard zone would be
associated with a different level of risk given
a nuclear accident.

TRANSPORTATION & TRANSMISSION HAZARDS

Example:
Transportation of Toxic Chemicals to
Waste Disposal Sites

Geometry:
Linear

Contoured Risk Surface:
Each point along the route would be associated
with a different probability of experiencing an
accident.

CONSUMPTION HAZARDS

Example:
Consumption of Untested Chemical
Additives in Food

Geometry:
Punctiform

Contoured Risk Surface:
Each person would be associated with a given
level of risk given his personal consumption
habits.

FIGURE 3   A Geometric Classification of Technological Hazards

potential impacts. Hazard taxonomies, such as the geometric one presented above, provide an indication of the most important characteristics of technological hazards. Litai's analysis of a life insurance company's data base yielded nine salient dimensions of variation of which the four most important were found to be: (1) volition; (2) severity; (3) origin; and (4) effect manifestation (Rasmussen 1981:135-138). The Clark University/Decision Research technology assessment group proposed five primary dimensions of variation. Their hazard taxonomy classifies hazards according to: (1) whether materials, energy, or information is released; (2) the extent of knowledge about the hazard; (3) whether the hazard is intentional or unintentional; (4) the hazard's catastrophic potential; and (5) public perceptions of risks and benefits. Grouping hazards according to these variables should "suggest generic ways of identifying and coping with hazards, thereby contributing to more coherent public policy" (Clark University/Decision Research 1979:6). Observations on these variables may be added to the hazard's geographic dimensions to create multi-dimensional hazard profiles. To fully understand the management implications of these typologies, however, we must first explore the concept of risk and its relationship to technological hazards.

2

Technological Hazards, Risks, and Benefits

What is the difference between hazards and risks? Although often used inter-changeably, these terms have taken on meanings which communicate slightly different ideas. A hazard is a negative outcome which may take such forms as loss of life, personal or transgenerational injury, property damage, diminished productivity, psychosocial stress, or ecosystem degradation. Smalley (1980:138) even speaks of "hazards to pleasurable uses" of the environment such as a waste disposal plant that diminishes the attractiveness of a nearby park. Risk, on the other hand, is the probability that a particular negative outcome or assemblage of negative outcomes (often spoken of as scenarios) will occur. Kaplan and Garrick (1981:12), in fact, define risk as "uncertainty plus damage." They express the relationship between hazard and risk as follows:

$$\text{Risk} = \frac{\text{Hazard}}{\text{Safeguard}}$$

What this equation means is that the probability of an event with deleterious consequences is directly proportional to the magnitude of the hazard and inversely proportional to the safeguards implemented to guard against it. At the very least, these safeguards may take the form of hazard awareness. If the dangers are more subtle or difficult to handle, formal education may be needed to encourage hazard-averting behaviors in the home or workplace, on the road, or in any other risk-charged environment. Safeguards may also be built into the technological system itself so that it operates more safely. As these safeguards are increased, according to the equation, the hazard is reduced to "acceptable" or "minimum threshold" levels, but seldom to zero.

The above model succinctly summarizes the steps in a risk assessment: (1) identification of the hazard; (2) estimation of the risks; and (3) selection of the safeguards. These steps may be completed by searching out answers to the questions presented in Figure 4. Note that two pathways (denoted by the vertical arrows) are provided to link the hazard with the safeguards; one includes the risk assessment step and the other bypasses it. The basic questions asked by the risk analyst are accompanied by a set of questions which may be asked by the geographer interested in bringing out the spatial dimensions of the risk analysis. The outline presented in this figure forms the basis for the remainder of the chapter.

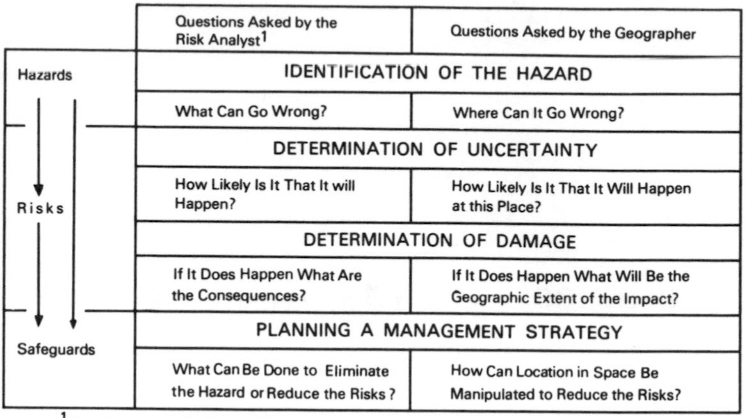

	Questions Asked by the Risk Analyst[1]	Questions Asked by the Geographer
Hazards	IDENTIFICATION OF THE HAZARD	
	What Can Go Wrong?	Where Can It Go Wrong?
	DETERMINATION OF UNCERTAINTY	
Risks	How Likely Is It That It will Happen?	How Likely Is It That It Will Happen at this Place?
	DETERMINATION OF DAMAGE	
	If It Does Happen What Are the Consequences?	If It Does Happen What Will Be the Geographic Extent of the Impact?
	PLANNING A MANAGEMENT STRATEGY	
Safeguards	What Can Be Done to Eliminate the Hazard or Reduce the Risks?	How Can Location in Space Be Manipulated to Reduce the Risks?

[1]Based in part on Kaplan and Garrick 1981: 13.

FIGURE 4   The Risk Analysis Process

Identification of the Hazard

There are two ways in which hazards may be identified. The first is through observation of a technology's negative consequences. This empirical approach to hazard identification looks back in time in an attempt to answer the question: What has gone wrong? It is an acceptable method so long as the negative externalities of a technological application are mild and have an observable legacy. The second way of identifying hazards is through the anticipation of a technology's consequences before they actually occur. This theoretical approach looks forward in time in an attempt to answer the question: What could go wrong? As the scale of the technology increases, as the effects are delayed in time, and as the consequences hint at catastrophe, the need to anticipate hazards before they arise becomes increasingly essential. Clark (1977:114) suggests that "the hazard of our present predicament is not an inability to predict the outcomes of our trials, but rather an inability (and unwillingness) to live with error." In many cases the prediction of hazards may not be an attainable goal.

Kates (1977a, 1977b, 1978) has grouped various methods of applied science commonly used to identify hazards into three categories: (1) screening, in which a standard procedure is used to classify items according to their hazard potential; (2) monitoring, in which observations of hazardous events or consequences are continuously recorded and analyzed; and (3) diagnosis, in which hazards are identified based on their symptoms or consequences. In the case of synthetic chemicals, the Kates model suggests that we could wait for the hazards of a particular chemical application to occur and then try to diagnose the cause. Once a specific chemical has been linked to a series of deleterious consequences, a monitoring system could be set up to provide an early warning of impending harm. To cope with highly toxic chemicals, a better strategy for hazard identification would be to develop a method to screen all chemicals for carcinogenic or mutagenic properties before they are used at all.

Screening is used by the Food and Drug Administration (FDA) to determine whether food additives may be marketed publicly. Federal law, in the Delaney clause, expressly forbids adding to food anything which causes cancer in animals. Nevertheless, as Johnson (1978) points out, some suspected carcinogens, particularly the coal-tar food dyes, are being used by the industry. One of these dyes, Red No. 2, was banned by the FDA in 1977 but the carcinogenicity of other coloring agents is still a subject of contention. In general, however, food additives and drugs are not marketable until their safety is established. Chemicals used outside the body, on the other hand, are often not subjected to rigorous screening. Children's sleepwear treated with Tris-BP, a flame-retardant, was on the market for more than five years before tests demonstrated its carcinogenicity in animal populations (Blum *et al.* 1978). Even more remarkable, however, is that it remained on the market two and a half years after studies first uncovered its mutagenic and carcinogenic properties (Regenstein 1982:282-284). Garments treated with Tris were finally banned from sale in the United States in 1977; unsold articles were shipped to markets abroad.

If screening is not employed to identify hazards, we may remain unaware of negative consequences until they appear in an exposed population and are diagnosed. Thalidomide, a sedative and sleeping agent, underwent three years of animal tests in West Germany (Lawless 1977:140-147; Sjostrom and Nilsson 1972), and was subsequently approved for over-the-counter sales in that country and several others in the 1950s. Thereafter, it was taken by many pregnant women to relieve morning sickness. Unfortunately, tests had not been conducted to detemine its effects on the developing fetus, an oversight which led to one of the great technological tragedies of the century. It was only through diagnosis of thousands of babies born with "flippers" instead of arms, hands, or feet, that thalidomide was discovered to have transgenerational consequences for pregnant women who took the drug during the first twenty to forty days of gestation.

While many food additivies, pesticides, and drugs have been nationally proscribed on the basis of screening and diagnosis, other harmful chemicals are tolerated in the environment and their levels are monitored to assure that concentrations do not exceed so many parts per million or billion. The Environmental Protection Agency establishes tolerance levels for pesticide residues in food, and the Department of Agriculture monitors animals coming to slaughterhouses for more than 50 chemicals and drugs. The USDA's sampling system assures that there is a 95 percent chance that at least one violation will be detected if the actual violation rate in the population is one percent or more. The EPA's National Human Monitoring Program samples the national population for residues of pesticides and toxic chemicals in blood, urine, and adipose tissue (Council on Environmental Quality 1983:57-67). Through programs such as these, government agencies are able to detect threatening environmental pollutants before they reach unacceptable levels. One of the problems with the monitoring approach, however, is that we often lack sufficient knowledge on which to base maximum allowable concentrations before a chemical substance enters the hazard category.

A comprehensive hazard theory would make it possible to predict hazardous outcomes of a technology's application. Kates (1977b:22-23) provides two examples of propositions that may lead to anticipatory hazard identification. First, the hazardousness of chemicals is directly proportional to their rarity in nature. Synthetic compounds which are non-existent in the natural world are likely to be the most toxic. Second, hazards occur at disjunctions of energy or material cycling or flow. As Turner (1978:1) has noted, "the kinds of energy which man now makes use of are inherently much more destructive than those which he has traditionally controlled." In line with this idea, a theoretical systems model has been proposed by Olson (1979) who suggests that hazards result when flows between the following four hierarchically arranged subsystems are not in equilibrium: (1) the circulation system of the human body; (2) the workplace; (3) the social economy which circulates goods, information, and money; and (4) the earth itself. The human body and the earth are seen as stable systems while the workplace and the social economy are seen as expansive. Because growth characterizes the economy and the workplace, velocities of exchange between them and the two stable systems increase and hazards result. For example, if demand for a product increases as part of an expanding social economy, pressure is put on the workplace to step up its output. Initially, this will mean a higher volume of material and energy flow through a plant of given capacity. To illustrate the type of hazards that result, Olson cites the oil refining industry which reduced down-time between 1968 and 1973 and thereby experienced more injuries, falls, and intoxications. As the workplace expands to accommodate increased demand, not only do accidents and occupational disease increase, but the workplace generates pressures on the earth's environmental systems in the forms of resource depletion and pollution. Olson suggests that in periods of rapid growth in the social economy and the workplace, we can expect higher rates of workplace accidents and environmental crises as a result of what might be termed the "negative externalities of growth." She then goes on to suggest that hazards also increase in geographically isolated systems and as a result of gravity which makes humankind "vulnerable to the vertical." Of the fatal work accidents in Quebec, for instance, one-third resulted from falls or falling objects. Although Olson was concerned primarily with explaining occupational accidents and diseases, her model holds promise for a holistic and potentially predictive theory for identifying a full range of hazards.

Estimation of the Risks

Once a hazard has been identified, probability theory is generally applied to assess the likelihood (1) that a hazardous event will occur, and (2) that particular consequences will result. Rasmussen (U.S. Nuclear Regulatory Commission 1975:9) provided an example of risk estimation using empirical data for automobile accidents which is updated here. In 1980 there were an estimated 17,777,000 persons involved in police-reported traffic accidents (U.S. Department of Transportation 1982:iii). One out of every 348 was a fatality. Since risk is the probability of a negative outcome (in this case death), we need to multiple the frequency of the event by its magnitude:

$$\frac{17{,}777{,}000 \text{ persons involved}}{\text{year}} \times \frac{1 \text{ person killed}}{348 \text{ persons involved}} = \frac{51{,}083 \text{ persons killed}}{\text{year}}$$

Reduced to the individual level, the probability of having been one of those killed in a police-reported motor vehicle accident was:

$$\frac{51{,}083 \text{ deaths/year}}{226{,}504{,}825 \text{ persons}} = \frac{2.255 \times 10^{-4} \text{ deaths}}{\text{person-year}}$$

If the above calculations for 1980 are representative of other years, this means that the chance of an individual being killed in a motor vehicle accident is one in 4,434. Deaths,

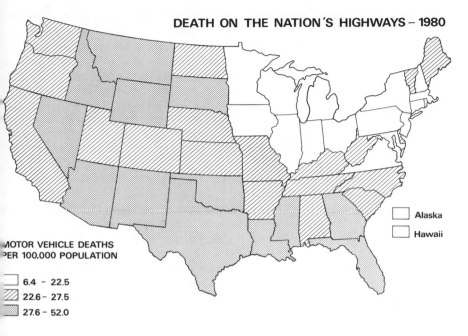

DEATH ON THE NATION'S HIGHWAYS – 1980

Alaska

Hawaii

MOTOR VEHICLE DEATHS
PER 100,000 POPULATION

6.4 – 22.5
22.6 – 27.5
27.6 – 52.0

FIGURE 5   Death on the Nation's Highways (National Safety
Council 1981)

however, are only one possible negative outcome of automobile accidents. Similar
calculations could be made for injuries, delayed deaths, and property damage.

These calculations are based on the total population of the nation in 1980.
Because this may not be the most appropriate base, the risk of death or injury is often
calculated per 100 million vehicle miles traveled or per 10,000 registered motor
vehicles. Furthermore, not everyone within a population is equally exposed to the risk
so that the population base is often standardized for age, family status, sex, and other
variables. In 1981, for instance, 35.8 per cent of the drivers involved in fatal accidents
were under age 25 (U.S. Department of Transportation 1983:i).

Location is another important determinant of the degree of risk to which an
individual is exposed. Death rates from motor vehicle accidents are mapped out by
state in Figure 5. Although the national motor vehicle death rate in 1980 was about 22.6
per 100,000 population, the range of death rates extended from only 6.4 per 100,000 in
the District of Columbia to 52.0 per 100,000 in Wyoming. Translating these death rates
into risk estimates and minding the caveat that not all fatalities in a state are residents of
that state, the risk of being killed on Wyoming's highways was 1 in 1,923, while it was
only 1 in 15,625 in D.C. The fatality pattern portrayed on the map suggests a strong
rural-urban dichotomy: Almost 2 out of 3 traffic deaths in 1980 occurred in rural areas
(National Safety Council 1981:41). The scale of the map hides much intrastate
variation between urban and rural areas, however. Even though more accidents do
occur in urban areas, lower speeds mean that accidents are more likely to be "property
damage only" or "injury only" accidents. Higher speeds on rural highways mean that
the accidents which do occur are more likely to involve fatalities.

If we use 1980 traffic fatality data to estimate the future probability of being killed in a motor vehicle accident we are employing a method of risk estimation known as extrapolation (Kates 1977a, 1978). Extrapolation forward in time from an historical record, backward in time from imagined events, and laterally by analogy to similar hazards are the three methods commonly used to assign probabilities to events and their outcomes. Predictions of natural diasters are usually based on an historical record: volcanic eruptions, 50-year floods, and tornado incidents, as well as traffic accidents, are assigned probabilities of occurrence by extrapolation forward in time. For some technological hazards, simple forward extrapolation may produce over-estimates of risk since technological systems are being constantly re-designed to minimize mishaps.

Extrapolation backward in time is used to assign probabilities to hazards which theory indicates could occur but which never have. An example would be the method-ologies used to assign probabilities to severe nuclear accidents. Extrapolation by analogy to similar circumstances may be illustrated by the collapse of the I-95 bridge over Connecticut's Mianus River in 1983, after which bridges of similar construction were immediately designated as high-risk links on the national highway system. Similarly, risks associated with LNG tankers are calculated by examining the safety records of ships which carry gasoline, liquified petroleum gas, and chlorine (Smalley 1980:143-144). The limitations of risk estimation by analogy are illustrated by Idaho's Teton Dam. Because none of the hundreds of earth-fill dams built by the Bureau of Reclamation had failed, the Teton Dam was thought to be equally as safe. Never-theless, in 1976 the dam broke, releasing a 17-mile long reservoir of water which killed eleven and left 15,000 families homeless (Laycock 1976).

In the total absence of a data base for extrapolation, Kates (1977a, 1978) identified only personal intuition and revelation by supernatural inspiration as other methods of risk estimation. History is replete with predictions of doom based on divine revelation. Risk estimation by personal intuition may yet become more amenable to scientific analysis if methodologies can be perfected for "pooling" the intuition of groups (Westcott 1968). Foster (1980:150-154) noted the amenability of the Delphi technique to hazard prediction and risk estimation. In the Delphi technique, a group of knowl-edgeable individuals is selected to anonymously complete an original, open-ended questionnaire and then react to feed-back about the results on an iterative basis through several rounds of questioning. The results may take the form of a list of future hazards, their likelihood of occurrence, predicted magnitude of impact, and ways in which government agencies should react. Because this procedure can be oper-ationalized without a formal data base, it is one of the best methods yet developed to peer into the long-term future.

Logic trees are sometimes employed in the extrapolation of risks in order to determine (1) the probability of an undersirable event taking place, in which case they are called *fault trees,* and (2) the probabilities of various outcomes given a hazardous event, in which case they are called *event trees.* Slovic and Fischhoff (1983:44) have presented a simple fault tree (Figure 6) to show how radioactive wastes could be released to the biosphere after being sealed in a salt bed. Rare events and tech-nologies without a track record are often analyzed using fault trees. Probabilities are typically assigned to each pathway in a fault tree so that the ultimate probability of a hazardous event occurring may be calculated. One of the difficulties, however, is underestimating these ultimate probabilities by failing to consider all of the possible

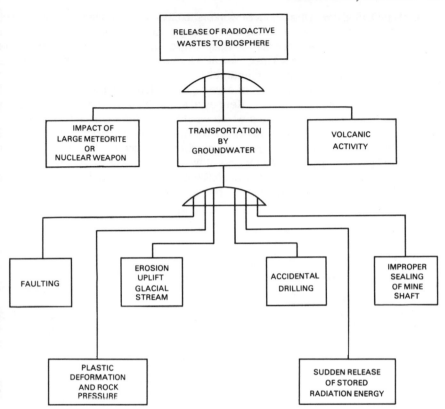

FIGURE 6 A Fault Tree Analysis of a Radioactive Waste Release (McGrath 1974, cited in Slovic and Fischhoff 1983:132)

pathways that could lead to a mishap. Pathways which are commonly omitted from these analyses include human error and sabotage, failure to anticipate changes external to the technological system, failure to conceptualize how a complex system functions as a whole, and failure to consider "common mode failures," that is, those that result from the same cause (Slovic and Fischhoff 1983:132). The diagram in Figure 6, for example, fails to include the pathway representing corrosion of the storage vessel by virtue of salt's propensity to chemically react with other substances. Many of these common ommissions are illustrated by the fire at Alabama's Browns Ferry nuclear power plant in 1975, an accident which was not anticipated, yet almost resulted in a core meltdown. The cause of the fire was human error: a technician was searcing for an air leak with a lighted candle while both nuclear reactors at the site were operating at full power. Once ignited the fire spread and proved to be more dangerous than originally thought. Not only did it burn through the electrical cables which controlled the primary core cooling system of one reactor but it also knocked out the cables which provided power to the back-up systems. According to Sheridan (1980:25), human errors "occur with alarming frequency in complex systems." At Japan's Tsuruga nuclear generating station near Osaka, it was the failure of workers to close a holding tank's valves which resulted in 40 tons of primary coolant water being discharged into the sea. This was Japan's most severe nuclear accident — again the result of human error. At Three Mile Island, on the other hand, it was a mechanical failure which triggered the human failure to close a valve, a combination of human and mechanical error.

Given the occurrence of a hazardous event, the various outcomes that may result are often organized into an event tree. Keeney *et al.* (1978) have constructed an event tree to project the number of deaths which would result from Liquified Natural Gas (LNG) tanker collision in the harbor at Port O'Connor, Texas, a possible site for an LNG receiving terminal (Figure 7). Some possible outcomes of such a collision show that the release of LNG might be followed by one of two types of events: (1) an immediate ignition of the vapors, or (2) the spread of LNG vapors over the ground until they come in contact with ignition source in a boat or home. Probabilities have been assigned to

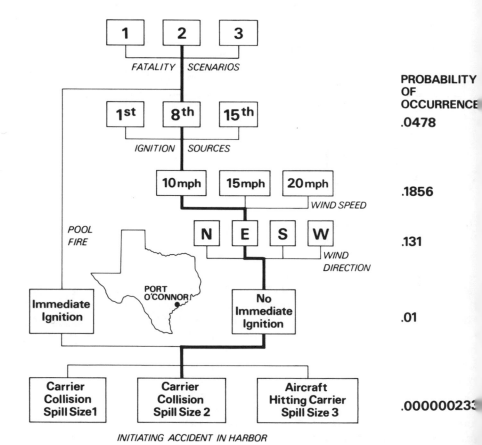

FIGURE 7   An Event Tree Analysis of a Liquefied Natural Gas Explosion (Redrafted from Keeney *et al.* by permission from *Technology Review*, copyright 1978)

each event along the hypothetical pathway which is shown by a heavy black line on the diagram. A probability of .01, for instance, means that there is only one chance in 100 that there would be no immediate ignition. If this chain of probabilities is multiplied together, it gives the chances that on a tourist-season weekday, one tank of LNG would be released and ignition delayed until the vapor had spread into a residential neighborhood. By multiplying these numbers, the ultimate probability may be calculated as 2.71 $\times$ 10^{-12}, or .00000000000271. By superimposing the hazard zone on a population density map of Port O'Connor, the authors calculated the annual risk per person under this scenario to be 2.5 $\times$ 10^{-11}.

The negative consequences of a technology may be vested on people, property, productivity, or natural systems. Of human consequences, some are easy to define: death and personal injury for example. Others are more difficult to assess: shortened lifespan, transgenerational injury (e.g., mothers given DES during their pregnancy to prevent miscarriage unknowingly exposed their daughters to a higher risk of vaginal cancer), and psychosocial stress. The process of technology assessment demands "a systematic study of the effects on all sectors of society that may occur when a technology is introduced, extended or modified, with special emphasis on any impacts that are unintended, indirect, or delayed" (Lawless 1977:5) The impact of a technology on the environment may be analyzed by employing the various methods of environmental impact assessment developed during the 1970s in response to the National Environmental Policy Act (Greenberg et al. 1978; Clark et al. 1980; Lee 1983). Porter et al. (1980) distinguished between a technology assessment (TA) and an environmental impact statement (EIS) by noting that the former is not site-specific and covers a broader range of possible ways to regulate the technology. EISs, on the other hand, are usually conducted to assess the impacts of a particular technology at a particular site; they are more "procedurally structured" in order to meet National Environmental Policy Act requirements. Both TAs and EISs are likely to project, perhaps with logic trees, the positive and negative consequences of a technology and the likelihood that they will occur. While technology assessments lay out the options for policy-makers, Kahn et al. (1976:168-169) colorfully illustrate the hazards of the methodology with the following scenario:

> Let us assume that U.S. authorities had made a TA study of the automobile in 1890. Assume also that this study came up with an accurate estimate that its use would eventually result in more than 50,000 people a year being killed and maybe a million injured. It seems clear that if this study had been persuasive, the automobile would never have been approved.

What is also clear, albeit missing from the foregoing scenario, is the necessity for including an estimation of the benefits as well as an enumeration of the risks in a technology assessment. The residents of Mount Desert Island, Maine, for instance, initially banned the automobile because it threatened to change the leisurely pace of island life. When a patient died because his car-less doctor could not reach him in time, the value of automobiles came to be appreciated and in 1913 the law banning their use was revoked (Rich 1970:293). The number of lives saved each year nationwide as a result of the time-space convergence brought about by the automobile may be difficult to enumerate, but it surely is much greater than the number of people killed in automobile accidents.

At the heart of every debate over technological risk is the question: How safe is safe enough? While impact analysis and probability theory may be applied to the calculation of risks, the decision on whether the risk is too high to accept cannot be made by scientists. Scientists can provide information for the decision-making process, perhaps even recommendations, but the ultimate choice of a technology rests with the individual, the corporation, or the government. If risk can be meaningfully translated into monetary units, benefit-cost analysis may be an appropriate method of decision-making. This methodology pits the advantages of a technology against the disadvantages in order to determine its acceptability:

$$\text{Acceptability} = f \left(\frac{\text{Benefits}}{\text{Positive Costs} + \text{Negative Costs}} \right)$$

In the above equation, costs have been parcelled out into two terms, positive costs and negative costs. The former are the expenditures which must be made in order to implement the technology and enjoy the benefits; the latter are the costs associated with the negative externalities of a technology's employment. Both costs are *numerically* positive. 'Positive' suggests costs willingly invested; 'negative' costs are unfortunate by-products of the technology. If the benefits outweigh the costs, the technology may be brought into production. If the same benefits can be realized from several different technologies, the one with the lowest costs will be chosen.

All too frequently the attempt is made to reduce the human costs to dollar amounts. Dardis (1980), for instance, calculated the value of a human life in the late 1970s to be between $257,000 and $295,000! This is often called the "human capital," "foregone earnings," or "deferred future earnings" approach to assigning a value to a life. During the past decade an alternative method has become more popular: The "willingness to pay" measure is calculated on the basis of how much the private or public sector is willing to pay (through, for example, investments in safety measures) to save human lives. Both approaches call to mind Schumacher's (1973:46) critical characterization of benefit-cost analysis: "a procedure by which the higher is reduced to the level of the lower and the priceless is given a price."

If the second term in the denominator of the above equation (costs associated with negative externalities) is the primary variable of interest, the benefit-cost analysis becomes a benefit-risk analysis:

$$\text{Acceptability} = f \left(\frac{\text{Benefits}}{\text{Risks}} \right)$$

If B/R>1, then the technology is acceptable
If B/R<1, then the technology is not acceptable

This means that the personal, corporate, or public decision to accept a technology is *directly proportional to the benefits* and *inversely proportional to the risks*. Decisions are made, however, on the basis of perceived risks and benefits which are not always the same as "scientifically" derived measures. Slovic *et al.* (1979) have shown, in fact, that for many technologies there is wide disparity between the way experts evaluate risk and the way laymen evaluate risk.

Two generalizations may be made on the basis of the above benefit-risk model. First, if there are no benefits to be gained, it does not matter how small the risks are, the technology will be rejected. Conversely, if there are no risks, even small benefits could bid the technology into use if it passes the benefit-cost test. Similarly, if B/R>1 the decision on whether to adopt the technology would revert back to the benefit-cost equation. Once the technology has passed the test of social acceptability (benefit-risk analysis), it then becomes a candidate for the test of economic viability (benefit-cost analysis). Second, if the level of benefits goes up, the degree of risk an individual or society will be willing to accept will also be higher. Conversely, as the level of benefits goes down, the level of acceptable risk will also decline. Starr (1969) was the first to validate this principle in his pioneering effort to quantify technological risk and social benefit. As an example, Foster (1980:17) noted that "the higher degree of employment and profit associated with a technology, the greater the socially acceptable risk."

In their attempt to quantify perceived risk, Slovic et al. (1980:202) asked a group of college students to rate the perceived risk and benefits of ninety hazardous activities, substances, and technologies. They found that the greater the perceived risks, the more risk adjustment was needed to make the hazards acceptable, and that the greater the perceived benefits the less risk adjustment was needed. In their assessment of the Three Mile Island accident the same authors (Slovic et al. 1982) observe that the only factor which is likely to broaden the base of public support for nuclear power in the short-run is a severe energy shortage. They note that the oil embargo of 1973-1974 brought public support for the controversial Alaska pipeline, off-shore drilling, and shale-oil development despite their enviromental risks. The foregoing equation suggests that as benefits are perceived to increase, greater risks are likely to be accepted. Barring an energy crisis, the authors see the need to restructure public perceptions of both risks and benefits. They do not believe that this is possible by bombarding the public with more information. Instead they see the path to public acceptability as dependent on three developments: a long-term safety record, a trusted regulatory system, and a clear appreciation of benefits.

There are serious obstacles in the way of employing benefit-risk analysis. First, public decisions about risk are made more difficult because people do not seem to perceive risk as a linear function of deaths, injuries, and damages. The public seems to "rely on individual heuristics which tend to systematically bias perceptions of risk" when compared with objective statistical data (Cole and Withey 1981). Starr (1969) was the first to quantify the notion that voluntary risks are accepted more willingly than involuntary risks despite objective measures of death and destruction. This finding has been validated by Slovic and Fischhoff (1983) and by Litai (Rasmussen 1981). Slovic and Fischhoff report a similar "double standard" for risks which are familiar, known, controllable, and immediate, all of which the public seems to be more willing to accept than risks which are new, unknown, not controllable, and delayed. Perhaps these dichotomies help to explain why pressure is mounting to increase the speed limit from its 55 mph ceiling despite the fact that a lower speed limit — even lower than 55 mph — could considerably reduce the risks of being killed in an automobile accident. Smalley (1980), however, draws our attention to the distinction between the involuntary risks to which we are resigned (natural disasters and disease) and the involuntary risks which are imposed on us by outside entities. Starr (1969) concluded, in fact, that the upper limit of acceptability of technological risk is set by the risks of disease. Factors such as volition, controllability, and familiarity distort the objective evaluations of risk derived from mortality and injury rates. One of the judgmental biases to show up in studies of how people evaluate risk is the tendency to judge an event as likely if it is easy to recall or envision (Tversky and Kahneman 1974). This seems to be why Slovic et al. (1979)

found deaths from accidents, homicides, fires, tornadoes, and floods to be over-estimated by the public while deaths from disease were underestimated. Research into the quantification of risk perceptions should aid in the benefit-risk analysis process.

Second, benefit-risk analysis is difficult because benefits and risks are measured in different units. Benefits are most often calculated in dollars or lives saved. Risks on the other hand are probabilities which are not easily transformed into common units of measure. How are social, psychological, and environmental costs to be transformed into dollars and cents? Moreover, benefits are virtually always tangible, while hazards are often intangible, at least in the short run. Holdren (1982) sees the temptation to depend on only those factors which can be quantified and to compare that which cannot be compared as two major criticisms of benefit-risk analysis. Moreover, hazards often occur neither immediately upon the development and use of a technology nor in ways which are obviously linked to its deployment. Even if similar units were used to evaluate benefits and risks, deciding on a discount rate to apply to hazards which are not immediate in their impacts offers a similar challenge (Schulze *et al.* 1981; Pearce 1983).

Third, defining benefits and risks depends on where you bound the technological system. The implicit rule followed by capitalistic enterprise is that you internalize under a corporate logo the benefits of a technology and externalize as many of the costs and hazardous outcomes as possible. Corporate responsibility for the waste materials and by-products which they generate has traditionally ended at the loading platform where a sub-contractor assumes responsibility for their disposal. Disposal sites have been chosen on the basis only of their visibility — the less conspicuous the better. The negative externalities are therefore vested on society as a whole rather than being internalized by the firm creating the wastes. In the case of nuclear power, Weinberg has neatly summarized the contrast between the public and private sector in a risk-benefit matrix (Figure 8). How many cells of the matrix should a risk-benefit analysis include? Pasqualetti (1983) argued that environmental impact statements for nuclear power plants are limited geographically to the local area, when, in fact, there are more distant and lesser known impacts.

Fourth, benefit-risk analysis is difficult because some technologies have neither been tried nor have a track record long enough to enable the empirical verification of probabilities. Determining the likelihood of a civilian nuclear accident is like working in a vacuum, as attested by the vociferous debate which ensued over the probability assigned by the Rasmussen report (U.S. Nuclear Regulatory Commission 1975) to a sequence of events leading to a core melt in a nuclear reactor: one chance in 20,000

	RISKS	**BENEFITS**
PRIVATE	Financial Loss	Profit
	Company Survival	Growth of Company
PUBLIC	Radiation Hazard	Abundant Energy
	Proliferation	

FIGURE 8   A Benefit-Risk Matrix of the Nuclear Power Industry (Weinberg 1981:7)

reactor-years of operation. With little or no empirical evidence, expert opinion is often the only recourse for risk estimation, but the opinions of experts may vary widely, as evidenced by the response of the Union of Concerned Scientists (1977) which criticized the Rasmussen report on methodological and ethical grounds. The UCC found the calculated probability of a core melt accident to be "biased toward a low estimate of the potential risk" because the scientists who prepared the report did not have an adequate data base on which to make such a calculation, failed to consider all types of accidents which could lead to a meltdown, minimized the likelihood of common mode failures and design deficiencies in safety equipment, and ignored the possibilities of aging, sabotage, terrorism, and earthquakes. In addition, the UCC also criticized the report's assessment of accident consequences. An internal review of the Reactor Safety Study, known as the Lewis report (U.S. Nuclear Regulatory Commission 1978b), and other reviews (Yellin 1976) were also critical of its methodology.

Fifth, there continues to be a rather vigorous debate on benefit-risk analysis over what level of risk a society should be willing to tolerate in order to enjoy the benefits of a particular technology. These questions cannot be answered by plugging values into sophisticated mathematical models; they can be answered only by social and political institutions. The risk assessment process is therefore seen by Cumming (1981:1) as occupying the "interface between science and the society that created it." Not even the independent variables (attributes of different technologies) which affect public opinion with respect to different technologies are known: over 50,000 people per year are killed in automobile accidents, while no one has been killed by a nuclear reactor accident. Yet, no one advocates eliminating the automobile, while a large segment of society advocates eliminating nuclear reactors. What is different about these two technologies which causes the public to evaluate them differently?

Sixth, the interjection of values compounds the problems of benefit-risk analysis. No technology is developed in a value-free environment; a technology is brought into being because it will enable the attainment of something of value to an individual, a firm, or society. Opinions vary widely over what is valuable, however. The nuclear power controversy may not be so much a dispute over the risk of the technology (though that is often the outward manifestation of the debate) but over the benefits, that is, over what nuclear technology stands for and the values it promotes. Similar debates have also surfaced over the benefits of Petri dish pregnancies, recombinant DNA, and weapons technology.

Seventh, higher-order consequences of a particular technology are often difficult to incorporate into the benefit-risk equation. First-order consequences — immediate impacts of a technology — may be beneficial but second- and higher-order consequences may be deleterious. During the decade of the 1970s, society became increasingly sensitized to the principles of ecology and the way in which a single event can be felt throughout an entire system over a long period of time. For example, over 1000 chemicals are used as pesticides. The first-order consequences of their use is the control of pests, a decided benefit to humankind. The fact that pesticides also affect non-target organisms and have a tendency to move up the food chain, as Kepone is now doing along the lower James River in Virginia (Terman and Huggett 1980), had been ignored in the comparative evaluation of risks. Similarly, the medicated feed ingested by cattle has increased productivity and profits but it is difficult to measure the second order consequences of medicated feed in the form of residues that accumulate in meat and the number of resistent pathogens that evolve as a result of their use. In the end, how does one place a value on the extinction of a species and the reduction in the planet's genetic resource base? Some benefits of plant and animal species preservation are economic, such as pharmaceutical development and improved food-crop hybrids, but others are more difficult to translate into monetary terms, such as the aesthetic value of species, the ethical rights of species to co-exist with humankind, and the value of maintaining the stability of a biosphere on which the human species depends for its survival (Eckholm 1978; Ehrlich 1981).

Finally, there continues to be a problem for benefit-risk analysis in linking deleterious results with their causes, particularly if they are disassociated in time and space. Before risks can be approximated, measured, or injected into a benefit-risk equation, they must be firmly established as hazardous outcomes of a particular technology. The primary factor which has held up policy implementation to reduce industrial emissions of sulfur oxides has been the unwillingness of business, industry, and government to admit that there may be a cause-and-effect relationship between industrial pollution in one locality and acid rain and snow hundreds of miles away. In another instance, an increased incidence of hypothyroidism (underactive thyroid glands) in the newborn downwind from the Three Mile Island reactor in the period after the accident suggests that the release of radioactive iodine in 1979 may have been responsible and should be dealt with as one of the hazards of nuclear power generation (Macleod 1981). No causal link, however, has been definitively established; thus, whether the costs of such an outbreak should legitimately be included in risk-benefit analyses of nuclear power remains open to debate. In a similar vein, defining the links between the multitude of cancers and the technological and natural environment is becoming one of the research thrusts of medical geographers in the 1980s (Greenberg *et al.* 1980; Mayer 1981; Glick 1982).

Planning a Management Strategy

Once a technology's consequences and risks have been determined, (or at least considered), individuals, corporations, and governments must decide (1) whether the technology is to be employed as if the consequences of its application did not matter; (2) whether it is to be completely eliminated; or (3) whether it is to be managed so as to reduce its negative outcomes. The first, or "do nothing" approach, is the easiest course to follow when the risks are external to the decision-maker, time-delayed in their impact, or non-catastrophic. It is also common to do nothing when it is beyond a decision-maker's economic or political limits to take action. Preston and colleagues (1983), for instance, examined a lower income lakefront neighborhood in Hamilton, Ontario, which was subject to both natural (flooding and storm damage) and technological (pollution and noise) hazards. They found that people perceived the environment as threatening but expected nevertheless to continue living there. Rather than taking action to reduce the hazards, the residents made a series of "cognitive adjustments" to relieve the "cognitive dissonance" of living in an admittedly hazardous area. They would deny the existence of some hazards, admit to a diminishing awareness of others, and accept the promises of technology to overcome still others at sometime in the future.

The second or "technology eliminating" approach to risk management is more rarely used at the aggregate decision-making level. Individuals, however, do have the power to accept or reject consumer technologies on the basis of their personal risks and benefits. At a different scale, the threat to the stratosphere's ozone layer, which screens out ultraviolet radiation, resulted in a ban on chlorofluoromethanes as aerosol propellants. Chemicals have also occasionally been banned from use but global bans are virtually nonexistent. The insecticide Kepone was so toxic it was banned from use in the United States, only to be manufactured in Hopewell, Virginia, for export to Central America and Poland. After World War II, the global application of DDT promised a pest-free future. In 1962, however, Rachel Carson detailed the peril of DDT in the biosphere. Finally, in 1972, DDT was banned from use in the United States. Because DDT continues to be used in the developing world at the rate of 10,000 tons per year (Holdgate *et al.* 1982:282), its progression up the food chain continues to pose a global threat as it becomes concentrated in the fatty tissues of birds and mammals. To completely eliminate DDT worldwide, however, would mean an increase in insect-born diseases such as malaria. The elimination of threatening technologies can be temporary too. The United States had not manufactured chemical weapons since 1969 but recently re-embarked on a campaign to supplement its strategic arsenal with binary chemical weapons to be produced in Pine Bluff, Arkansas.

Now we seem to be entering an era of "better living through biology." What chemical engineers were to the past, biological engineers may be to the future. Biogenetic manipulation, which has already been through a decade of ups and downs (Krimsky 1982), may again become one of the most vociferously debated technological hazards of the late twentieth century and may result in the elimination of certain life-altering technologies (King 1980; Rifkin 1983). In the previous chapter it was pointed out that the most toxic of chemicals are synthetic, that is, they do not occur in nature. Does this suggest that synthetic life forms can be expected to be any less dangerous than the pathogens already found in the natural world? Cambridge, Massachusetts, and Ann Arbor, Michigan, are but two communities where DNA research being conducted in their midsts has been debated in public. The hazard of eliminating one technology after another, however, is that all benefits will also be eliminated unless a safe alternative can be implemented as an acceptable substitute. Does foregoing some of these benefits mean major sacrifices in quality-of-life?

In a related category are those technologies which may be headed down the road to elimination even if a conscious, outright decision has not been made to reject them. LNG terminals may not have been eliminated at the national level, but if cities and states continue to question their safety and prohibit them, the long-term prospects for growth in the industry appear bleak (Ahern 1980). Despite its economic benefits, the U.S. nuclear power industry may also be headed in this direction. The promises of the nuclear industry in the 1950s (safe, clean, cheap, and abundant electricity) are seldom heard today. The *de facto* moratorium on nuclear power plant licensing, time delays of 12 to 15 years from inception to completion of a plant, higher construction costs and cost overruns, and the number of citizen protests and law-suits suggest a less than promising future for nuclear power.

The third, or "hazard management" approach, depends on safeguards to eliminate or reduce the hazardous consequences of a technology's application (Fischhoff *et al.* 1978). Just as the outcome of natural hazard research has been the generation of resource-management strategies, the outcome of technological hazard research should be the generation of technology-management strategies. According to O'Riordan (1979:261), risk management may be thought of as "a search for the safest route between social benefit and social loss." The management of technology has traditionally been accomplished through the economic sector. Innovations were bid into production because they enabled the more efficient use of resources. The new thrust in technology management, however, seeks to manage technology not through the processes of supply and demand but through the social system and collective decision-making processes. In California, the public decision-making process has greatly reduced the probability that an LNG terminal will be built along the California coast and assured that if it is built it will be equipped with all requisite safeguards.

Preventing the hazardous consequences of technology may take the form of (1) isolating the technological system, either geographically or operationally; (2) regulating and perhaps limiting the use of the technology; or (3) modifying the technological system to minimize negative externalities. These options may be illustrated by the location of some noxious facilities away from population centers, by placing nuclear reactors inside containment vessels, by regulating the use of synthetic food additives and drugs, and by installing scrubbers on coal-fired power plants. Even the 'bomb' has been modified to limit its impact: the radius of destruction of a neutron bomb is only one-fifth as large as that of a conventional atomic weapon.

Hazard management strategies are also influenced by perceptions of hazards. If people perceive themselves to be in control of a technology, for instance, they may resist management strategies: almost nine out of ten Americans regularly fail to use automobile seat belts. If, on the other hand, people perceive the negative consequences of a technology to be beyond their control, especially if accompanied by a catastrophic potential, they are likely to demand very stringent safeguards. Litai (Rasmussen 1981:135-138) found that people are willing to accept a risk of death 100 times higher for voluntary hazards than for involuntary hazards, 30 times higher for both ordinary (as opposed to catastrophic) and delayed-effect (as opposed to immediate-effect) hazards, and 20 times higher for natural hazards than for man-made hazards. These generalizations may also indicate how willing people are to pay for hazard management. Reissland and Harries (1979:809), for instance, assert that a community is willing to spend at least ten times as much to save the life of a named individual as to save a "statistical" life.

A prerequisite to decision-making in the control of hazardous technologies requires the identification of "the causal chain of hazard evolution." Once this chain has been delineated, control points may be targeted and evaluated for policy-directed

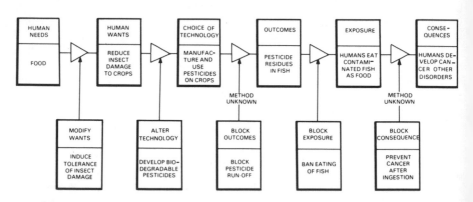

FIGURE 9   An Analysis of Pesticide Management Strategies
(Fischhoff *et al.* 1978:19)

intervention. Figure 9 presents the case of pesticide contamination plugged into a generic model developed by Fischhoff and colleagues (1978:18-19). This model is meant to provide a "common language" for the analysis of hazard management strategies and a basis for evaluation. Mitigating the hazardous consequences of technology may take the form of remedying damage which has already occurred, or stopping short the chain of hazard evolution once it has been set into motion. Treating cancers that result from exposure to radiation would be an example of the first and administering a blocking agent to prevent the thyroid gland from absorbing radioactive iodine would be an example of the second.

Hohenemser and collaborators (1983) have taken this general model and used it to structure a dodecadimensional hazard typology comprising a list of characteristics that can be enumerated for all hazards. These twelve dimensions of variation are listed in Table 4. They have been grouped according to the step in the causal model of hazard evolution to which they apply. The attributes of a hazard, as specified in the table, influence the way a hazard is perceived and the way it should be managed. The authors

TABLE 4 COMPONENTS OF A MULTIDIMENSIONAL HAZARD PROFILE

Stage in the Model Of Hazard Evolution	Hazard Descriptor	Explanation
Choice of Technology	Intentionality	Degree to which technology is intended to harm.
Release (Of Materials or Energy)	Spatial Extent	Maximum distance over which event has a significant impact.
	Concentration	Concentration of released energy or materials relative to natural background
	Persistence	Time over which the release remains a significant threat to humans.
	Recurrence	Mean time interval between releases above a minimum significant level.
Exposure (To Materials or Energy)	Population at Risk	Number of people potentially exposed to the hazard.
	Delay of Consequences	Delay time between exposure to the hazard release and the occurrence of consequences.
Consequences (Human and Biological)	Annual Mortality	Average annual deaths due to the hazard
	Maximum Potentially Killed	Maximum credible number of deaths in a single event.
	Transgenerational	Number of future generations at risk from the hazard.
	Potential Non-Human Mortality	Maximum potential non-human mortality
	Experienced Non-Human Mortality	Non-human mortality that has actually been experienced.

Source: Hohenemser *et al.* 1983:379. Reprinted by permssion of *Science*, copyright 1983.

see their typology as a method for building "multidimensional hazard profiles" so that new hazards can be compared with already existing ones in order to identify unexpected problems, suggest management strategies, and red-flag hazards that demand special attention because of extreme scores in one or more areas.

The hope of hazard management is that the benefits of a technology may be enjoyed and the risks controlled. Since 1968, carbon monoxide emissions from new automobiles have been reduced by more than 90 percent due to catalytic converters and modified engine designs (Council on Environmental Quality 1982:27). Overall, in fact, the nation's air and water are cleaner today than they were on Earth Day 1970 which ushered in the era of environmental awareness. Similarly, levels of toxic substances in coastal waters and high seas are now lower than they were ten years ago (United Nations Environment Program 1982). The outright rejection of DDT and certain other pesticides has worked to reduce the concentration of DDT in human adipose tissue over the past five years; it has also stimulated the development of "third generation pesticides," natural pest control agents such as bacteria, specific viruses, and sex confusants (Council on Environmental Quality 1983:65,105). Sometimes the solution becomes part of the problem, however. Tris-BP, for instance, was used to treat sleepwear because it was a fire retardant; asbestos was used to coat schoolroom ceilings and walls for the same reason. Both were found to be carcinogenic. Macgill and Snowball (1983) suggest that these and other hazard management strategies be evaluated on the basis of the following criteria: (1) Effectiveness — Is there an improvement in safety or protection? (2) Efficiency — Is the improvement achieved by the most parsimonious means? (3) Equity — Is there an equal distribution of costs and benefits? The equity criterion raises a host of geographic questions which are treated in the next chapter on the regional dimensions of benefit-risk analysis.

3

The Regional Dimensions of Benefit-Risk Analysis

Three questions must be answered in an equitable benefit-risk analysis: Who is going to realize the benefits? Who is going to experience the risks? Where is each group located? All too often, benefit-risk analyses do not adequately treat the final question of the trilogy. Instead, risk analysts operate in a spatial vacuum and it is tacitly assumed that those who experience the benefits also experience the risks. Occasionally this is true, but as the scale of technology increases there is most often a spatial disassociation between benefits and risks. The four possible outcomes of a geographic benefit-risk analysis are diagrammed in Figure 10, which shows that the region receiving the maximum benefit of a technology may be different from the region experiencing the maximum risk. "A very heavy burden of risk on one group, while another group gains most of the benefits, is clearly inequitable," noted Callahan (U.S. Congress 1980:130), so that a moral analysis must accompany any benefit-risk analysis. He suggests that more attention should be devoted to the question of how benefits and risks *ought* to be distributed. An example of how the increasing scale of technology imposes heavy burdens on populations which have no control over the risks is the 1978 wreck of the supertanker *Amoco Cadiz* off the coast of France which resulted in the largest oil spill in history. Even though the coastal communities of Brittany had no control over the technology and had received none of the benefits, it was their coast which suffered severe environmental damage. For more than a hundred miles, plant life was decimated, thousands of birds were killed, and shellfish operations lost millions. Another hazard with such negative externalities is noise pollution. As Harvey et al. (1979:264) note: "Social, psychological, and economic costs resulting from aircraft noise are imposed upon certain residents in the vicinity of airports who receive no direct benefits."

In Figure 10 the first possibility (A) is one in which the zone of benefits is coexistensive with the zone of risk. Many consumer technologies fall into this category: Those who are availing themselves of a particular product are those who are taking the risk of experiencing negative consequences. Holdren (1982:75) has pointed out that one of the advantages of renewable energy resources is that the decentralization of energy production would mean that those who were using the energy would also be subject to the hazards and potential environmental effects (Holdren et al. 1980).

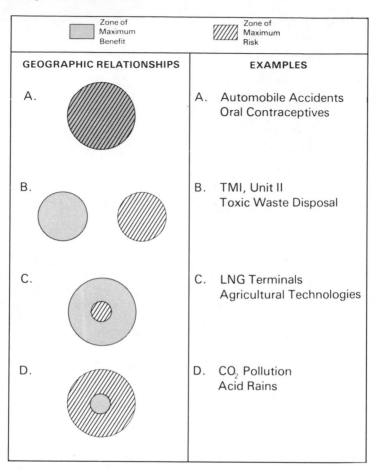

FIGURE 10   Technological Hazards: Regional
Dimensions

Centralized fossil fuel and nuclear plants, on the other hand, tend to "export major environmental burdens" and thus are neither equitable nor efficient. Perhaps an indicator of "appropriate technology" is how closely the geography of benefits and risks conforms to geometric model **A** in the diagram.

The second possibility (**B**) is a complete geographic separation of benefits and risks. Three Mile Island's Unit II reactor, for instance, was initiated by the Jersey Central Power and Light Company in the 1960s because of a growing demand for energy in the greater New York City region. Oyster Creek, New Jersey, was the site where construction began. When the organized crime syndicate in the state co-opted the labor union working on the project and presented Jersey Central with an extortion demand, the company decided to seek another location beyond the syndicate's grasp. The site finally settled upon was in a sister company's Pennsylvania territory, an island in the

Susquehanna River (Martin 1980:1-10). As a result of this process, the risks of nuclear power came to be imposed on south central Pennsylvania, while the benefits were intended for the growing consumer market in New Jersey. Nearby communities receive only a few jobs and tax dollars in return for their risk acceptance. Pijawka and Chalmers (1983), in fact, have documented the almost insignificant economic impacts of nuclear generating facilities on local communities around Three Mile Island and eleven other sites. They also report that in Pennsylvania, unlike many other states, utilities pay taxes to the state rather than the local community. The state redistributes these revenues on a per capita basis to all communities, thus minimizing Metropolitan Edison's potential positive impact on the local public sector. Even Starr and Whipple (1981:9) of the pro-industry Electric Power Research Institute have recognized this geographic inequity by noting that "nuclear power plant neighbors bear the public risks without direct compensation, except for property tax benefits." In Pennsylvania, the local community does not even receive property taxes.

The third and fourth possibilities (**C** and **D**) in Figure 10 apply to technologies where there is a discrepancy in scale between the zone of benefits and the zone of risks. All of New England, for instance, benefits from the LNG which is off-loaded in Boston, but neighborhoods around the harbor experience the maximum risk (Smith 1976). By the same token, the world markets may benefit from the application of agricultural technologies in certain regions. Yet, the risks associated with salinization, waterlogging, the use of pesticides and fertilizers are bestowed on the producing regions alone. The benefits of paper, steel, and petrochemical industries are felt throughout the consuming world but the negative externalities are concentrated at the production site. At the other end of the spectrum are the technologies whose risks are global but whose benefits are localized. The industrial nations, for instance, benefit from the burning of fossil fuels, while the entire world may suffer the consequences of a global warming trend as a result of the "greenhouse effect." (Mingst 1982; Revelle 1982; U.S. Environmental Protection Agency 1983). Likewise the high smokestacks that rid a locality of sulfur oxides by pumping them high into the atmosphere turn the problem of acid rains into a regional dilemma (Likens and Bumann 1974; Hendrey 1981; National Academy of Sciences 1983; Rhodes and Middleton 1983; Wetson and Foster 1983).

A temporal dimension has been added to the foregoing analysis in Figure 11. Time is portrayed along the vertical axis; the horizontal axis represents an assembly of places. The circled letters represent technological benefits and hazards, which are represented as both immediate and local for technology **A**. Dodge (1982), for instance, found that homes with gas stoves had a higher concentration of carbon dioxide and that children from these homes had higher rates of cough but did not seem to suffer any more severe or long-term lung damage. Technology **B** represents a delayed effect hazard but one which occurs in the same place where the benefits were enjoyed. In Bellevue, Ohio, sinkholes were used for public and private waste disposal beginning in 1872. The effects were not immediately noticeable but over the years the town's groundwater supply became completely contaminated and in the early 1960s some wells still gushed raw sewage (Epstein et al. 1982:302). Technology **C** represents a geographic disassociation between risks and benefits. As discussed above, the risks of the Three Mile Island Unit II reactor were realized in 1979 and the negative consequences vested on south central Pennsylvania, only a few months after the reactor went commercial to meet the demand for electricity in New Jersey.

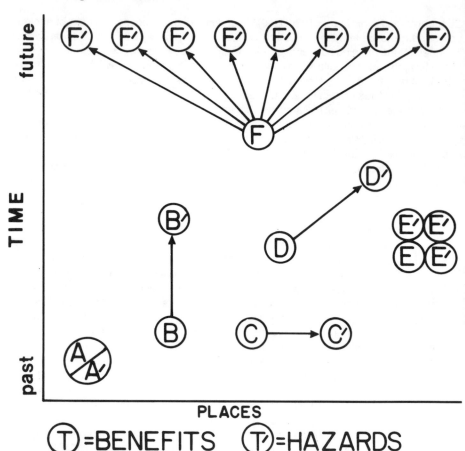

FIGURE 11   Time-Space Dimensions of Technology's
Benefits and Hazards

The case of toxic chemical disposal at places which are far from the production sites may serve as an example of Technology **D** where the benefits have been bestowed on one place at one time, while the associated hazards have been shifted to another place to be experienced at some time in the future when the disposal drums begin to leak. Technology **E** represents a contagious diffusion of the impacts in both space and time to surrounding regions as well as the place of origin. On-site disposal of sanitary wastes provides such an example. Technology **F** may represent the burning of fossil fuels which bestows its benefits on the industrialized regions leading to a global warming trend that will sometime in the future affect the entire world. Placement of some technologies within the matrix is more problematic: The negative impacts of an all-out nuclear war would be worldwide, but where are the places that benefit from such a technological confrontation? For any technology, how much of this matrix a benefit-risk analysis should cover is open to debate on both philosophical and practical grounds.

A Context for Understanding Locational Conflict

The foregoing time-space models of benefits and risks are useful in three ways: (1) they aid in the understanding of support for and opposition to new technologies; (2) they suggest the scale (individual, community, regional, national, global) at which decision-making powers should be vested; and (3) they aid in answering questions about who should bear the economic and social costs associated with a technology's negative side effects. All three considerations are essential to the understanding of locational conflicts that develop over technology's place in the world. For example, toxic wastes, whether chemical or nuclear, most often are not disposed of in the region where they were generated or used, as the people of New Jersey's pine barrens, the town of Montague in Michigan, and the residents of at least a hundred dioxin-contaminated communities in Missouri can attest. Because of this geographic dissassociation it has become exceedingly difficult to find socially acceptable locations for the disposal of toxic materials.

In 1976, the Alpena area of Michigan was selected by the federal government as a possible disposal site for high-level radioactive wastes. Commending the area were its sparse population, its isolation from major urban centers, and its position atop deep saltbed formations. The site selection criteria did not include the attitudes and perceptions of local residents, as it seldom does in such cases, nor did it include any input from state and local authoriies. A survey of the local population two years later found that over 90 percent were opposed to serving as a "dumping ground" for the nation's nuclear industry and 42 percent said they would consider moving if the Alpena site were chosen. In contrast, only 47 percent of the same population said they were opposed to nuclear power (Brunn et al. 1980). These findings illustrate the conflict between a society which desires the benefits of a technology and the necessity for concentrating the risks on one particular site. Once a site is designated a possibility for hazardous waste disposal, the response is usually the same: *Not In My Backyard!* (Jakimo and Bupp 1978).

Almost every place is somebody's backyard, however. With the development of nuclear technology after World War II, the oceans became a dumping ground for radioactive wastes, probably because they were perceived to be beyond everybody's backyard, well suited to the task because of their volume, and (best of all) uninhabited. Between 1949 and 1970, when the practice stopped, the United States dumped an estimated 95,000 curies of radioactive wastes at six dump sites less than 200 miles from shore on both coasts (Boehmer-Christiansen 1983). Studies of these waste sites have shown that radioactivity has not remained confined to the disposal site. Farallon Island, off the coast of California, has been contaminated with plutonium from a nearby dumpsite, for instance. What was originally perceived to be nobody's backyard now seems to have become everybody's backyard, and the practice of ocean dumping has been severely criticized. The London Dumping Convention of 1972 totally prohibits high-level radioactive waste dumping but does permit the disposal of low-level wastes under controlled conditions. Since the mid-1970s, however, only the United Kingdom, Switzerland, Belgium, and the Netherlands have dumped low-level radioactive wastes in the ocean. In 1983, during a meeting held to amend the London Dumping Convention, nineteen nations voted for a two-year moratorium on low-level radioactive waste disposal in the oceans; only six nations voted against the moratorium. These six included Japan, a nation without a vast territorial base which plans to begin ocean dumping by the late 1980s, and the United States, which has announced consideration of similar intentions (Curtis 1983). As the United States reconsiders ocean dumping for radioactive wastes, locational conflicts will surely spread to the oceans which now seem to have a constituency of their own. Even if they lack human inhabitants, numerous environmental and scientific organizations will argue against the practice of dropping lethal materials into the global commons.

With respect to noxious facilities such as dump sites, landfills, missile silos, and polluting industries, who should participate in the locational decision-making process and who should hold final site-selection power? If the decision-making process is an open one, as it should be in a democratic society, participants are likely to include consumers, industrialists, environmentalists, community organizations, legislators, and multitudes of public agencies. Each group will perceive the benefits and risks differently because of the limits which they place on their own benefit-risk analysis. Corporate analysts are likely to emphasize benefits and risks to the corporation; environmentalists, the benefits and risks to the natural environment; and legislators are likely to factor in benefits and risks to their own political future. The limits may be geographic as well. Local groups are likely to limit the enumeration of positive and negative effects to the local area, while the federal government will see the necessity for broadening the base of the benefit-risk equation presented in the last chapter. *Where* you bound the system in cataloging "numerator" and "denominator" events is of critical importance in balancing the equation and in understanding the conflicts which arise over technology's potential impact. In the case of the Alpena dump site, how does one evaluate the benefit-risk equation when the benefits are largely national, while the risks are largely local? Can the numerator and denominator of the equation represent two different scales? Another example of the concentration of risk on one population in order for society at large to benefit is the above-ground nuclear weapons testing at the Nevada Test site in the 1950s. Residents of Nevada and Utah were not warned of any negative side effects and many were even invited to watch the detonation from a distance. Today the federal government refuses to admit any linkage between the tests and the higher incidence of leukemia, miscarriages, and birth defects downwind.

An increasing number of potential locational conflicts over the siting of nuclear power plants, disposal of radioactive wastes, and regulations regarding uranium mining have found their way onto statewide ballots. Since 1976, twelve states have placed nuclear referenda before their electorates. Similar referenda have been held in a number of cities and smaller communities in the United States. At least one foreign country, Austria (1979), has given its electorate an opportunity to decide the fate of its domestic nuclear power industry.

At the state level, the issues have varied, as has the phrasing of questions appearing on the ballot. A list of these state referenda appears in Table 5. The most common issues appear to be whether there should be legislative approval required for new power plant construction and whether tighter safety restrictions and monitoring are needed. Voters were seldom asked directly whether they favored or opposed nuclear power. The percentages in the table identify the magnitude of the anti-nuclear vote regardless of how the specific questions were worded. Voters who favored prior legislative approval of new plant construction, tighter monitoring, and approval of nuclear waste disposal were construed as "anti-nuclear" even though there may be many among them who do not fit the radical connotation of this term. The overall record of opposition to nuclear power is mixed. Of the sixteen votes, in only five instances did the anti-nuclear voters prevail. The seven 1976 referenda signaled strong support for nuclear technology as a source of present and future energy needs (Wenner and

TABLE 5   STATEWIDE REFERENDA ON NUCLEAR ISSUES

Year	State	Issue	Percent Yes
1976	Arizona	Legislative Approval of Any Nuclear Power Facility	33
1976	California	Legislative Approval for Licensing Nuclear Power Plants	33
1976	Colorado	Legislative Approval of Nuclear Plant Construction or Modification	29
1976	Montana	Restrictions and Monitoring of Plant Construction	44
1976	Ohio	Legislative Approval of Nuclear Power Plants and Related Facilities	32
1976	Oregon	Legislative Approval Required for Nuclear Power Plant Licensing	42
1976	Washington	Legislative Approval for Future Nuclear Power Facilities	33
1978	Montana	Strict Restrictions on Plant Construction	66
1980	Montana	Forbidding Disposal of Radioactive Waste	51
1980	Maine	Prohibit Generation of Electrical Power by Nuclear Fission	41
1980	Missouri	Restricting Plant Construction	39
1980	Oregon	Plant Licensing and Waste Disposal Require Voter Approval	54
1980	South Dakota	Regulates Uranium Mining, Plant Construction, and Disposal of Nuclear Waste	48
1982	Idaho	Legislative Approval Required Before Endorsing Construction of Nuclear Power Facilities	60
1982	Maine	To Close the State's Only Nuclear Power Plant	44
1982	Massachusetts	Referendum Required Before Any New Low-Level Nuclear Waste Site or Power Plant Could be Established	67

Manfred 1978). In six of the seven states there was a two-to-one margin in favor of nuclear power, an understandable outcome during this era of rising energy costs and desire for "energy independence." In environmentally conscious states like California, Oregon, and Washington, nuclear power was still favored by a sizeable majority of the electorate.

The referenda held since the Three Mile Island accident in 1979 evidence greater opposition to nuclear developments. In four of the eight states, anti-nuclear votes were successful. Montana, Oregon, Idaho, and Massachusetts supported measures that called for legislative or voter approval before new plants could be licensed or waste disposal facilities established. Montana voters also narrowly approved a measure that banned the disposal of radioactive wastes from outside the state. South Dakota voters rejected a referendum that called for the regulation of uranium mining, plant construction, and the disposal of nuclear waste. Only in Maine were the voters asked whether an operating nuclear reactor should be shut down; the vote was negative.

A number of village, county, and city electorates have also voted on nuclear power issues, including several counties in the Three Mile Island area which will be discussed later. In 1977, voters in Bakersfield, California went on record as opposing the building of a nuclear power plant in their community. The largest city that has held a referendum on a nuclear power issue is Austin, Texas. At issue was whether the city should sell its 16 percent share in South Texas Project Nuclear (STPN). The first vote was held in April 1979, immediately after the TMI accident. The referendum failed (49 percent) then, but was successful in 1981 (58 percent). Health and safety reasons were important to Austin voters but so were the higher utility costs that were associated with remaining in the STPN. The referendum passed in 1981 in part because the turnout was larger (by nearly 10 percent). The strongest support came from areas with low incomes, large proportions of the population below the poverty level, low median housing values, and large numbers of adults who had not finished high school. University of Texas students and the poor were most instrumental in garnering support to ensure passage of the referendum.

Hazard Zones as Contoured Risk Surfaces: Real and Perceived

One of the problems in delineating the hazard zones of many technologies is the absence of an historical record on which to base risk analyses. The measurement of "actual" risk therefore becomes a matter for the "experts" working with fault trees and computer models. Ideally, in planning for technological disasters, public perceptions of the risks ought to be commensurate with the reality of the threat. The public should neither over-estimate nor under-estimate the severity of the hazard. Unfortunately, in the absence of expert consensus on the threat posed by nuclear power installations, toxic chemical disposal sites, or offshore oil drilling, the perception of the magnitude of the hazard or the configuration of the hazard zone rarely conforms with official estimations. Perceptions of the technological landscape are in a constant state of flux. Translating technological threats into geographical patterns should result in a hazard management strategy that builds on public perceptions and behavioral intentions.

It goes without saying that the perceptual hazard zones around nuclear power plants nationwide have changed since 1979 and one wonders what methodologies could be developed to map these changing surfaces. In 1974, regardless of the reality of risk around TMI, no hazard zone seemed to exist in the minds of the local people. Until the accident, the hazard zone was considered to be co-extensive with the Low Population Zone (LPZ), a distance of 2-3 miles from the plant. In 1979, however, the perceptual hazard zone around TMI changed dramatically. Suddenly, a faltering nuclear reactor presented a threat to the local population, their property, and their long-term future. After the accident a new contoured risk surface materialized around the Metropolitan Edison facility. TMI is but one island in a nationwide archipelago, however, and around each nuclear power site in the U.S., local residents and the experts began to revise their estimates of risks and to modify their conceptions of the hazard zones. The Sandia Laboratories (1982) report to the Nuclear Regulatory Commission forces a further revision of popular and expert concepts of risk surfaces around nuclear installations. On the basis of available probabilistic risk assessments, the study assumed the chances of a severe core-melt accident to be 1 in 100,000 per reactor-year. Although at first glance the risk seems miniscule, this figure did little to allay fears of a reactor accident. If there are 100 reactors in operation, the chances of an accident occurring are increased to 1 in 1000 per year (.0001) so that over a twenty year time span, the chances of a "worst case" accident somewhere in the United States would be .02 or 2 percent (.0001 x 20 years). From figures contained in the report, it was calculated that as many as 102,000 early deaths (more than at any other site) could occur within a year of a "worst case" accident at the Salem nuclear station in New Jersey. What will this series of risk estimates do to the perceptual risk surfaces around each of our nuclear installations and how will the planning regions be adjusted to reflect these changes?

One year after the Three Mile Island accident, seventy-six residents in the TMI area responded to a survey conducted by the authors which contained a map of south central Pennsylvania. On that map they were asked to "shade in the area you believe was severely impacted by the TMI accident, considering both distance and direction from the plant." These seventy different perceptual surfaces (six did not fill out the map) were amalgamated into a composite contour map of the impact area by superimposing a 41 x 33 grid over each respondent's map and calculating the percentage of times each cell was shaded. The result is displayed as an isoline map with lines connecting points of equal perceptual severity (Figure 12). Not unexpectedly, the most severely

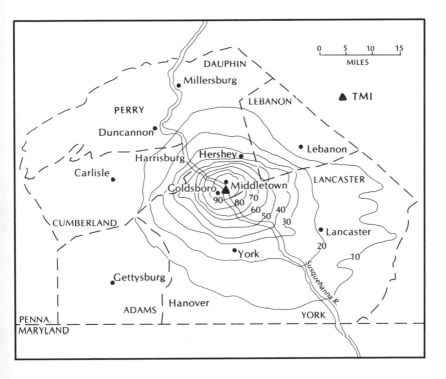

FIGURE 12   The Perceived Zone of Impact Around Three Mile Island

affected area was perceived to be immediately adjacent to the plant itself with a decided distance-decay effect determining the contours of the perceptual region. Consciousness of direction among the respondents is revealed by the elongation of the impact area downwind and downstream from the plant. Northwest of the installation the impact zone was perceived to extend only 15 miles, while to the southeast the 10 percent line extended 30 miles.

In contrast to the zone of perceived impact around Three Mile Island, the dosimetric map of the area as presented by the U.S. Nuclear Regulatory Commission (1979) indicates a minimum of actual risk from the somatic impact of radioactive fallout. Nevertheless, the psychological fallout from the accident has had lingering effects on the nearby population. Houts and colleagues (Houts *et al.* 1980; Houts and Goldhaber 1981) followed the symptoms of distress arising from the accident over an 18 month period. In comparing the population closer to the plant with a control group 40-55 miles away, they found several significant distance-decay relationships. The population within 15 miles of the reactor was significantly more upset both immediately after the accident and 9 months later. Only after 18 months had passed were there no significant differences in the percentage of people reporting that they were extremely or moderately upset. Similarly, it took 18 months for there to be no significant decline with increasing distance in the proportion of people reporting specific physical and behavioral stress symptoms. In contrast, a question which asked respondents how severe a threat TMI is to their families' safety continued to show a statistically significant difference in the proportion of people perceiving TMI to be a severe or very severe threat 18 months after the accident. Houts and collaborators (1980) believed that one of the reasons for this long-term impact, in sharp contrast to the lack of direct harm resulting from the accident, was that the crisis lacked a sense of resolution. Unlike the attitudes toward TMI held by the local residents before 1979 when a hazard was not even acknowledged, the residents today are acutely aware of being within the hazard zone of a nuclear reactor.

Three years after the accident at Three Mile Island the public reaction against nuclear power was reflected in a referendum on which much of the local population voted. The referendum asked voters in three counties: "Do you favor restarting TMI Unit I which was not involved in the accident on March 28, 1979?" As Harrisburg's *Sunday Patriot-News* (May 16, 1982:A4) editorialized, "The referendum is particularly significant because, for the first time, there will be a measure of the willingness of a population to accept an operating nuclear power plant back into its midst after suffering through the shock and fright of a major nuclear accident."

The results of this referendum are displayed in Figure 13 which shows the percentage of the population voting against the restart of the undamaged reactor. The map provides a glimpse of the public's perception of the risk of nuclear power, concern about safety, and resentment about the slow and clumsy clean-up of Unit 2. Voters in these three counties rejected the restart by a 2-to-1 margin but there was still a considerable amount of variation in voting patterns. Several factors help to explain these patterns as they appear on the map. First, there seems to be a friends-and-neighbors effect which may explain why the area closest to the plant did not vote as forcefully against the referendum as did municipalities slightly farther away. Second, beyond the "inner ring" there seems to be a distance-decay effect which explains the low percentage of no votes farthest from the plant in western Cumberland County and in eastern Lebanon County. Third, there seems to be a self-interest effect which may explain why Lebanon County, where 91 percent of the population are served by

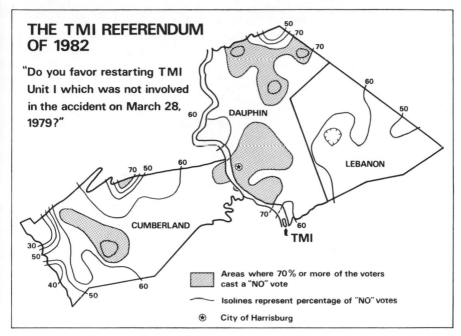

FIGURE 13   Mapping a Referendum Vote to Evaluate the Acceptability of Nuclear Power

Metropolitan Edison, voted more favorably on the restart that did Dauphin and Cumberland Counties where only 5 percent and 7 percent respectively are served by Metropolitan Edison. The self-interest effect may also explain why the Commissioners of York and Lancaster Counties, which are almost completely served by Met-Ed, did not allow the issue to appear on their ballots. Fourth, there seems to be an effect of urban-rural extremes. Much of the Harrisburg urbanized area and two of the most rural areas, one in the mountains of northern Dauphin County and one in west-central Cumberland County, voted the most resoundingly against the restart. Despite the vote of no confidence, however, the utility was given the go-ahead to restart the undamaged reactor at Three Mile Island.

Perception vs. Reality: Management Implications

Many technologies are perceived to be more threatening than they really are; other technologies present risks which are perceived to be minimal when, in fact, they are substantial. Ideally, public perceptions of the technological risk should be commensurate with the reality of the threat. Arriving at such an equilibrium position through an open risk assessment process should provide a solid basis for intelligent decision-making about risk reduction. Approaching risk perception from a geographic perspective, there is a need to bring the perception of hazard zones into conformity with their actual geometric configuration. This concept is diagrammed in Figure 14 which

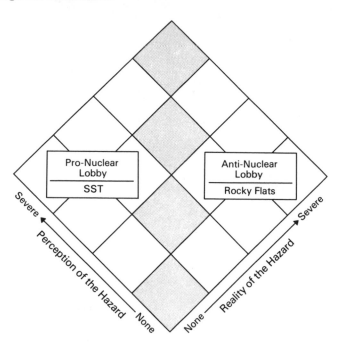

FIGURE 14　　Perception versus Reality: Public
Policy Implications

portrays the perception of the hazard along one axis and the reality of the hazard along the other. All hazards could be placed within the matrix according to (1) the reality of the threat that they pose, and (2) the public perception of the threat. In practice, however, the former is often unknown (some would even say unknowable for many hazards) and the latter is often subject to considerable debate. As an idealized model, however, the matrix helps us to put technological hazards into perspective and suggests a direction for hazard management. First, management initiatives should be directed toward the diagonal so that people's perceptions of the hazard are in line with the actual risks, and so that their perceptions of the hazard zone match the reality of its configuration. Second, management initiatives should be directed toward the origin, the point at which both the real and perceived risk have been reduced to zero.

　　The left half of the matrix is the setting for many technologies which pose minimal actual risk but are perceived as threats to society. In other words, they are considered to be more threatening than they really are. The pro-nuclear activists and the nuclear industry itself believe that nuclear power plants currently present very little risk to the public and that the threat is perceived by many to be greater than it really is. The Rasmussen report (U.S. Nuclear Regulatory Commission 1975) estimated the chances of an average citizen being killed in a nuclear accident to be about the same as the chances of being killed by a falling meteorite. Cohen (1981) argues that nuclear power's risks are well understood, minimal, and nothing compared to the risks of a coal-fired plant. He even suggests that nuclear power should be viewed as a means of "cleansing the earth of radioactivity" by getting rid of uranium deposits and thus eliminating the formation of radon gas in the natural environment. Kopecki (1981) sees the nuclear option as the way to meet the world's increased demand for energy and the most expeditious means of reducing air and water pollution and relieving pressure on the earth's exhaustible resources. Clearly, these advantages are not perceived as redeeming virtues by a large segment of the public. Another example is provided by a technology which made possible high-altitude, high-speed flight. The SST's impact on the earth's ozone layer was perceived in the early 1970s to be so severe that the technology was rejected by the U.S. Congress. Some recent evidence suggests, however, that high-flying SSTs may actually encourage the formation of ozone rather than breaking it down (Toth 1981).

These two cases illustrate the risks of operating to the left of the diagonal: worthwhile technologies may be rejected because they are perceived to be more dangerous than they really are. Alternatively, resources may be pumped into hazard management activities in an effort to reduce public exposure to a level of risk that has been overestimated. The first course of action would mean foregoing the benefits of a technology that is actually benign. The second course would mean squandering resources on reducing a non-existent threat, when they could be used elsewhere in meeting higher-order community needs.

The right half of the matrix (Figure 14) is the setting for technologies that present a threat which is either not perceived or perceived to be less severe than it actually is. The anti-nuclear lobby, for instance, would have us believe that the reality of the nuclear hazard is severe and that the industry's assessment of the risk does not measure up to the reality of the threat. They believe that catastrophic risk cannot be adequately expressed in probabilistic terms; that the more nuclear reactors there are in operation the greater the likelihood of an accident somewhere; that the entire nuclear fuel cycle (including mining, transportation, reprocessing, and waste disposal) is intolerably hazardous; that the magnitude, irreversible and long-term effects of the radiation hazard are so severe that society should reject the technology outright; that sabotage, terrorism, and natural disasters can be neither completely controlled nor predicted; that centralized, capital-intensive energy production is incompatible with democratic values because it concentrates power in the hands of an elite; and that renewable resources can just as easily be developed as the nuclear option to supply us with energy.

The case of the Rocky Flats nuclear weapon plant northwest of Denver, Colorado, provides an example of a hazard zone which was not perceived to exist until the delayed-effect hazards made themselves known almost two decades after the plant began operations in 1953. Johnson (1981) used the plutonium content of the soil around the nuclear installation as a measure of the cumulative exposure, 1953-1971. Figure 15 presents Johnson's contoured risk surface calibrated in millicuries of plutonium per square kilometer. Exhaust from the plant's smokestack was the primary source of plutonium so that the configuration of the hazard zone is a function of wind direction and distance from the source. For each band in the diagram, Johnson calculated the percentage of excess cancer deaths, 1969-1971. He found a higher incidence of all cancers near the plant and a decided distance decay effect with decreasing concentrations of plutonium. The spectrum of cancers associated with the survivors of Hiroshima and Nagasaki were also the ones most in evidence around Rocky Flats — leukemia, lymphoma, myeloma, and cancer of the lung, thyroid, breast, esophogus, stomach and colon. Had the residents around Rocky Flats recognized the true configuration of the contoured risk surface, measures might have been taken to further reduce the release of radionuclides from the smokestack or to inform the population of the risks of living in the hazard zone.

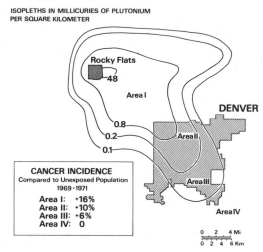

THE IMPACT ZONE AROUND A NUCLEAR
WEAPONS PLANT

ISOPLETHS IN MILLICURIES OF PLUTONIUM
PER SQUARE KILOMETER

Rocky Flats
48
Area I

DENVER

0.8
0.2   Area II
0.1

CANCER INCIDENCE
Compared to Unexposed Population
1969-1971
Area I: +16%
Area II: +10%
Area III: +6%
Area IV: 0

Area III

Area IV

0   2   4 Mi
0 2 4 6 Km

FIGURE 15   The Impact Zone Around the Rocky Flats Nuclear
Weapons Plant (Redrafted from Johnson 1981:177)

Another example of life in an unperceived hazard zone is provided by the famed incidence of mercury poisoning, Minamata disease, in Japan. A chemical company had been using mercury as a catalyst and dumping it in a nearby bay. By 1956 the symptoms of mercury poisoning had reached epidemic proportions in the local seafood-eating population. Nevertheless, the company kept evidence of the cause-and-effect relationship between industrial mercury and contaminated seafood secret for almost a decade (Harris and Hohenemser 1978). Over a hundred deaths and hundreds of illnesses occurred before the bay was finally closed to fishing in 1973 (Lawless 1977:251, 257). Only gradually did the people around Minamata bay become aware of the hazard zone in which they lived; and only gradually were hazard mitigation strategies implemented.

These examples illustrate the risks of operating to the right of the diagonal in Figure 14: populations may be exposed to hazards without knowing it; or, they may experience the symptoms and not understand the causes or relate them to environmental conditions. Geographically, someone may unwittingly be living in a hazard zone and not realize it until the risk has been realized.

The fact that some groups would place nuclear power on one side of the diagonal and other groups on the opposite side illustrates the problems of assessing the reality of risk, particularly for low probability/high consequence events, for delayed-effect hazards, and for technologies with which we have no record of experience. If we knew that the risks of nuclear power plant accidents were higher than we have estimated, the public sector could exercise its authority to modify the technology or alleviate its hazardous outcomes. On the other hand, if we knew that the risks of nuclear power were lower than we have estimated, educational programs could be implemented to bring people's perceptions of the hazard into line with reality. In the absence of consensus on the methods and results of risks assessment, however, public policy can only be the subject of continuing controversy.

4

Risk Mosaics

Technological hazards are unevenly distributed in both time and space. At any scale, from global to local, it is possible to define high-risk and low-risk environments. The geometrical technological hazard zone typology presented earlier, consisting of points, lines, areas, and volumes, may be thought of as the basis for defining the spatial distribution of risk. Technologies generate risk surfaces; the amalgamation of these surfaces makes it possible to delineate the total risk mosaic. The risk of being killed or injured in a motor vehicle accident, for instance, is easily calculated as a function of geographical location. Not only is it possible to recognize high-risk individuals but also high-risk environments. Specific intersections, bridges, entrance ramps, and stretches of highway may be targeted as high-risk areas and managed accordingly.

How people perceive their place within this risk mosaic determines many facets of human behavior, including personal risk management strategies. Perception depends on an intuitive understanding of both background risk and single-hazard risk: understanding the nature of the total risk mosaic is as important as understanding the risk of an individual technology or activity. It has been asserted that the greater one perceives the background risk to be, the less important a single hazard becomes. Starr (1969; *et al.* 1976) has suggested that the "psychological yardstick" for evaluating the acceptability of a particular risk is the average mortality due to disease for the entire population. In other words, the public seems to be willing to accept the risks of a technology or activity so long as it does not subject the population as a whole to a mortality rate which exceeds the disease mortality rate. Similarly, the natural hazard mortality rate seems to establish the lower bound of acceptable risk; that is, risks which are lower are not even considered in the risk analysis process, whether at the personal or societal level.

In understanding the total risk mosaic, it is also important to realize that many hazards which affect the same place may combine synergistically as well as additively and that the presence of one hazard may increase the risk of negative consequences from another. Thus, the holistic risk environment is more than the sum of its parts. In addition, risks that accrue to some individuals in a particular place may not accrue to other individuals: risks may be irregularly distributed in social as well as geographic space. For instance, the risk of deleterious consequences from toxic chemical exposure decreases with age; at highest risk is the fetus. On the other hand, some risks increase with age, either as a result of the aging process itself or because of prolonged exposure to harmful technologies or their by-products. By the same token there are hazards of childhood and hazards of adolescence, as well as hazards of gender, of being poor, of being handicapped, and of being employed in particular jobs. The Consumer Product Safety Commission estimates that 125,000 children are injured yearly in toy-related accidents. Among adolescents, one of the leading killers is motor vehicle accidents. While 15-24 year olds make up only 18 percent of the population, they accounted for 34 percent of the total motor vehicle fatalities in 1981 (U.S. Department of Transportation 1983:3). Childhood lead poisoning, particularly among the urban poor, is another example of a technological hazard which is neatly patterned in both social and geographic space. In a classic example of institutionalized discrimination, the black population in Houston, Texas has been forced by the dominant society to accept more than their share of the negative impacts on health and environmental quality generated by the city's solid waste disposal sites. Bullard (1983) found that incinerators and landfills in Houston were located primarily in black neighborhoods and that black children were more likely to attend schools close to solid waste dumps; both patterns which are likely to occur nationwide. Hazards, such as solid waste dumps, are often visible features of the cultural landscape and attempts to mitigate hazards for specific demographic, social, and occupational groups are often reflected on the landscape and in interior designs as well. Witness, for instance, the structure of children's playgrounds, of streetscapes where people must mix with automobiles, and of elderly apartment complexes.

The Globalization of Hazards

The global risk mosaic comprises both the sum of place-specific risks and the risks that are truly global in scope, such as increases in temperature as a result of the build-up of CO_2 in the atmosphere (Bach 1979; Revelle 1982; U.S. Environmental Protection Agency 1983), the depletion of ozone in the stratosphere (Karl 1980; Pyle and Derwent 1980; Isaksen 1981), and the contamination of the world ocean (Goldberg 1981; Kamlet 1981). Some technological disasters may be initially localized in impact but may have long-term consequences of global magnitude. One example is the diffusion of bioaccumulative toxic chemicals throughout the global food chain. Mercury and DDT have been reported in fish and wildlife caught on all seven continents. Another example is the uncertainty of thermonuclear war which would be targeted on particular population nodes and military targets but which would negatively affect the human population and the environment of the entire world.

The scale of impact of technological disasters has been increasing with the scale of technology and the widespread adoption of new, albeit untested, innovations. The larger the scale of potential impact, the more difficult it is to measure risk quantitatively. In the first place, as Groenwald (Stallen: 1980) observes, numerical measures of risk at the "macro-scale" are inadequate because of the enormous potential for harm should the risk come to pass. Given any probability of global destruction greater than zero, risks at the macro-scale are catastrophic sooner or later. In the second place, there have been few experiences with truly global disasters so that a quantification of risks becomes a matter of expert conjecture.

From the perspective of technological risk assessment, a sharp contrast divides the world's developing countries from the developed. The developing countries are technology-poor; the benefits of technology are therefore eagerly sought with little regard for the risks. That American multinational corporations spend less than half as much on pollution control overseas as they do in the United States (Berman 1978:182) illustrates an attitude toward technology's negative externalities that characterized American society one hundred years ago. The International Labor Organization has attributed the doubling and tripling of fatal accidents in the workplaces of developing countries to the adverse effects of the transfer of technology ("Occupational Hazards" 1983). Another example is provided by the international asbestos industry which has located the most hazardous steps in the production process in foreign countries with weak occupational health laws, such as Mexico and Brazil, where asbestos production sky-rocketed during the 1970s. Shue (1981:586) sees the shifting of negative externalities onto workers in the less developed countries as "extending the profitable life of physically harmful technology." Why should safety and health standards for workers in U.S.-controlled foreign plants be less demanding than those which apply to plants located in the United States?

The developed countries have passed through a technological transition in which standards of living, health, and well-being have expanded exponentially with advances in technology. Further improvements in well-being are now less responsive to technological advances with the result that inhabitants of the developed nations are questioning the immediate and long-term risks of technologies before accepting them as part of the cultural milieu. Nevertheless, in these same countries the "hazards of technology continue partially unchecked, affecting particularly the chronic causes of death that currently account for 85 percent of mortality in the United States" (Harris *et al.* 1978:38).

The global economic system has been able to take advantage of this division between developed and developing countries in a world where political decision-making is highly fragmented while economic systems have globalized. Corporations of the developed countries have effected a slow migration of high-risk technology and its output to the Third World. As the market for nuclear reactors among nations of the developed world declines, for instance, vendors seem more interested in marketing nuclear technologies in the less developed countries where safety expectations are less stringent. Even though fission reactors are often cited as "inappropriate technology" within the context of Third World development needs, the Soviets, in cooperation with the Finns, intend to market a small "turnkey" nuclear reactor in these nations (Egan and Arungu-Olende 1980).

Less developed countries have even been asked to accept the risks of technology without any of the benefits. A Colorado company, for example, offered Sierra Leone $25 million for permission to dump millions of tons of hazardous chemical wastes from the United States (Richards 1980). In addition, many North American and European firms have been looking to the Third World as a market for goods that are too hazardous to be sold elsewhere. Half a million infant pacifiers which were implicated in causing deaths by choking and banned from the U.S. market, were exported to Third World countries (Regenstein 1982:215). It has been estimated that in 1982 these countries bought over a billion dollars worth of defective goods from the United States (Islam 1983:50). Jacob Scherr, a Natural Resources Defense Council attorney, enumerated the following additional examples during hearings before the House of Representatives (U.S. Congress 1978:42-77). First, pesticide poisonings of farm workers have become a major problem in many developing countries as a result of exports from the developed world. Leptophos, a pesticide never registered in the U.S. for domestic use, was sold to Egypt where it resulted in death and illness among farmers and the deaths of over 1000 water buffalo. Second, in Colombia an outbreak of miscarriages and birth defects was linked to a herbicide similar to Agent Orange which continued to be exported from the U.S. even after it was prohibited for domestic use. Third, an antibiotic drug which is strictly limited to life-threatening infections in the U.S. because of serious side-effects, was marketed in Latin America by U.S. firms for tonsilitis, bronchitis, and even the common cold. Fourth, millions of dollars worth of sleepwear treated with the fire-retardant Tris-BP were exported to Third World countries after they had been with-drawn from the American market due to possible carcinogenic side-effects. To combat such international "corporate crime" the International Organization of Consumer Unions established a network in 1982 to warn member nations about the products that are being peddled worldwide even though they have been banned in the country of manufacture. Prior to instituting a comprehensive network, the IOCU had set up the Pesticide Action Network, the Baby Food Action Network, and Health Action International (with a focus on pharmaceuticals) to accomplish more specific hazard-management goals.

The Global Risk Mosaic: Nuclear War

Thermonuclear war represents the ultimate in technological catastrophe (Fiske *et al.* 1983; Peterson 1983). Since 1947, the most well-known estimation of the risk of such a confrontation has been the *Bulletin of the Atomic Scientists'* clock, a symbol of "the threat of nuclear doomsday" as judged by the nuclear experts. It is currently set at three minutes to midnight, the closest it has been in thirty years. As the superpowers return to a cold war arms race and as the number of nations with the bomb increases, the likelihood of nuclear war escalates. At present, six nations acknowledge having nuclear weapons: the United States, the U.S.S.R., the United Kingdom, France, China, and India. These nations, the years when they detonated their first nuclear bomb, and other potential bomb-building nations are depicted in Figure 16. Members of the nuclear club already include the world's four most populous nations and may soon include several of the world's smaller but most belligerent states. Israel and South Africa, in fact, may already have tested nuclear weapons, while Pakistan, Iraq, Egypt, Libya, Brazil, and Argentina are but a few of the nations which could probably build nuclear weapons now or within a decade. Many of these nations are in the Middle East, southern Africa, and the Indian sub-continent, three of the world's tinderboxes. The risk of war is real in each region, as it is between the superpowers, and the probability of a war becoming a nuclear confrontation grows as fast as membership in the nuclear club.

The effects of "multiple nuclear weapons detonations" were the subject of a study done under the aegis of the National Academy of Sciences/National Research Council (1975). Their report is probably the most exhaustive scientific study done to date on a nuclear war's potential impacts upon natural atmospheric, aquatic, and terrestrial systems; that is, on the non-human dimensions of a hypothetical nuclear exchange. Participants in the assessment of global effects were deliberately restricted to special-ists in the natural sciences. Had social scientists and humanists been included there would undoubtedly have been discussion of basic human needs, medical and mental health care, community reorganization, basic coping and survival strategies, and other issues such as those portrayed in the much-publicized ABC-television movie, *The Day After*. Also excluded from the report was an assessment of the immediate effects of the bombs in target areas, subjects which are treated in some detail by Glasstone and

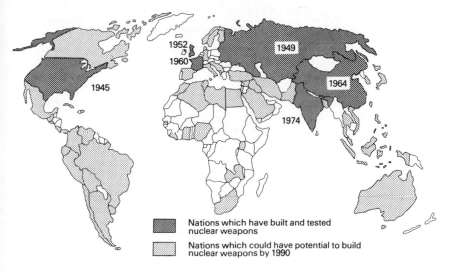

FIGURE 16   Nations With Nuclear Weapons Potential (Based on Congressional Research Service 1975:66-68; U.S. Arms Control Disarmament Agency 1976:220-221)

Dolan (1977), the Office of Technology Assessment (1979), and Openshaw and Steadman (1982, 1983a, 1983b). To exclude the immediate and local effects is to exclude a major part of the damage, since about 90 percent of the energy in a nuclear explosion is released in less than one millionth of a second, much in the form of heat and shock waves which devastate the target area (U.S. Arms Control and Disarmament Agency 1975:7). What the National Academy of Sciences report underscores is that the impact of a nuclear war will not be confined to target areas—the remaining ten percent of the energy released may be responsible for severe, long-term, global consequences. The report's projections are based on the following assumptions: that the detonations would likely occur in developed countries of the northern hemisphere, that the upper limits of the detonations would not exceed 10,000 megatons, and that weapons launched would arrive at their designated targets and would not veer off course. The findings of the National Academy of Sciences investigation are summarized in Table 6, which outlines the impacts on atmospheric systems, terrestrial ecosystems, and human beings in terms of somatic and genetic effects. The conclusions were that the biosphere and humankind would survive, that the direct effects of the detonation on non-target nations would be limited to thirty years, and that the indirect effects might be felt for a longer period of time.

No part of the world would be spared in the aftermath of even a limited nuclear confrontation. Just as the intensive nuclear testing between 1945 and 1971 reveals its global impact in Figure 17, similar fallout regions would materialize after a nuclear weapons exchange between nations in the northern hemisphere. The map shows the areas of highest cumulative concentration of Strontium-90, a radioactive isotope with a fairly long half-life of 28 years; it is based on an analysis of soil samples collected between 1965 and 1967. One of the patterns worthy of note is that the region of highest concentration covers one of the richest agricultural regions of the world, the "bread

TABLE 6   THE LONG-TERM EFFECTS OF NUCLEAR WAR

ATMOSPHERIC EFFECTS

1   Average cumulative fallout of strontium-90 of approximately 1 Ci/km^2 in the middle latitudes of the northern hemisphere, but "hot spots" two-to-three times more intense within this zone.

2   Injection of five-to-fifty times the background level of nitrogen oxide into the stratosphere, possibly resulting in a 30-70 percent drop in the northern hemisphere's ozone column and a 20-40 percent drop in the southern hemisphere's.

3   Temperature reduction of a few percent within one-to-three years as a result of dust injected in the atmosphere (about the same volume as the Krakatoa blast in 1883).

4   Most severe impact on climate in the high latitudes and least severe impact in the low latitudes.

NATURAL TERRESTRIAL ECOSYSTEMS

1   Plant life irradiated by 2 rads above background radiation but no widespread, long-term effects on plant ecosystems or on animal populations' survival.

2   Increase in ultraviolet radiation due to ozone depletion which might be fatal to many species of plants and the ecosystems of which they are a part.

MANAGED TERRESTRIAL ECOSYSTEMS

1   Long-term, worldwide contamination of cows' milk, but no long-term radiation damage to crops or animals.

2   No genetic changes in crops or animals but possibility of new strains of virulent plant pathogens which could lead to disease epidemics that would spread globally.

3   Significant impact on world agriculture which would necessitate adjustments in dietary patterns.

4   Shift in the northern limits of food production south in the northern hemisphere and shortened growing season in marginal areas.

AQUATIC ENVIRONMENT

1   No significant, observable effects on human populations, even those consuming contaminated seafoods, as a result of radionuclide concentrations in sea water.

2   Possibility of irreversible injury to aquatic species sensitive to ultraviolet radiation.

3   Reduction in the average temperature of the aquatic environment which would not significantly affect most species but which would result in a reduced geographic distribution of some.

SOMATIC EFFECTS ON HUMANS

1　Additional 4 rem exposure over 30 years which would yield an increase of about 2 percent in the spontaneous cancer death rate.

2　Radiation-induced anomalies in developing fetuses at time the detonations would occur.

3　Increase in skin cancer incidence of about 10 percent in the middle latitudes as a result of increased ultraviolet radiation.

4　Severe sunburn in the temperate zones and snow blindness in northern regions as a result of increased ultraviolet radiation.

GENETIC EFFECTS ON HUMANS

1　An increase in genetic diseases of 0.2 to 2 percent in the first post-detonation generation.

2　Total genetic effects extending more than 30 generations into the future.

Source: National Academy of Sciences 1975:5-17.

basket" of the United States and Canada. Considering nuclear detonations' impact on global climate, a subject which recent evidence suggests may have been under-estimated, food production in the middle latitudes may be endangered not only by radioactive fallout but also by a severe cooling trend that would shift southward the northern boundaries of crop production.

The recognition that the continued build-up of huge arsenals of nuclear weapons by the United States and the U.S.S.R. is enhancing the likelihood of large scale destruction has led various local, national, church, civic, conservation, and peace groups to push for passage of "nuclear freeze" proposals. These proposals, much like the following, have appeared on public ballots and before governing bodies nationwide in various forms.

> To improve the national and international security, the United States and the Soviet Union should stop the nuclear arms race. Specifically, they should adopt a mutual freeze on the testing, production, and deployment of nuclear weapons, and of missiles and new aircraft designed to deliver nuclear weapons. This is an essential, verifiable, first step toward lessening the risk of nuclear war and reducing the nuclear arsenals.

Statements in support of a nuclear freeze have come from national citizens' lobbies, major Protestant denominations, and bishops of the Roman Catholic church. Supporters include those of varying ages, socio-economic status, party identification, and political ideologies. One poll found that over 80 percent of the American population support a halt to the testing, production, and installation of nuclear weapons on the part of both the U.S. and the U.S.S.R. (Sussman 1983). The freeze resolution has been voted on at the local level nationwide and by late 1982, 371 communities in 32 states had passed resolutions supporting the freeze. This figure did not include the 44 town

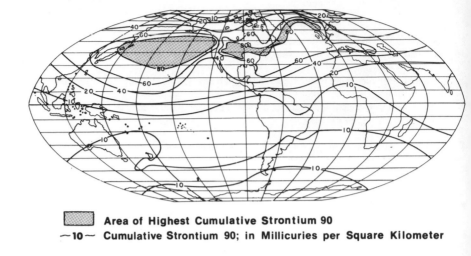

Area of Highest Cumulative Strontium 90
~10~ **Cumulative Strontium 90; in Millicuries per Square Kilometer**

FIGURE 17   Cumulative Concentration of Strontium-90 (Redrafted from Hardy 1968:586 by permission from
Nature, Vol. 219, p. 586; copyright © 1968 Mac-
millan Journals Limited)

meetings in five New England states or the 51 county boards in 17 additional states tha
went on record supporting the resolution. New Jersey, New York, Connecticut, Penn
sylvania, Ohio, and Iowa had the largest number of communities supporting the freeze
with small numbers in California, Oregon, Maine, Washington, Illinois, Minnesota
Maryland, Connecticut, and Massachusetts.

Statewide referenda on the freeze issue were on the ballot in the District o
Columbia and the following nine states in 1982: Arizona, California, Massachusetts
Michigan, Montana, New Jersey, North Dakota, Oregon, and Rhode Island. The
resolution passed in D.C. and all states except Arizona. California voters only narrowl
approved the resolution. It has been estimated that 10.8 million voters out of 18 millio
expressed a preference on the freeze issue during the general election (Stengel 1982)
Voters in two states, Montana and Colorado, also had an opportunity to vote on issu
directly related to the arms build-up. Montana voters went on record against basing M
missiles within their borders, a resolution that may be impossible to implement bt
which sends a definite message to military planners in Washington. Colorado voters b
a large margin defeated a proposed constitutional amendment that would have move
or closed the Rocky Flats nuclear weapons plant near Denver. Voters in Denver di
approve a nuclear freeze resolution, however. In short, the 1982 referenda were clearl
in support of a mutual and verifiable freeze.

With local and state support for the nuclear freeze resolution apparent in man
parts of the country, supporters took their case to the House of Representatives, th
same body which in 1982 had missed passing the resolution by only two votes. Afte
much delay engendered by opponents trying to weaken the language, the resolutio
passed on May 4, 1983, by a wide margin: 278 to 149. The issue enjoyed bipartisa
support; 218 Democrats were joined by 60 Republicans. The opposition was 10
Republicans and 43 Democrats. Some Congressional delegations, including thos

from California, Massachusetts, Michigan, Minnesota, New Jersey, North Carolina, Pennsylvania, Washington, and Wisconsin, were unanimous or nearly so in support. Opponents constituted a small but important part of the delegations from Florida, Illinois, Maryland, New York, Ohio, and Texas. Alabama, Georgia, Louisiana, South Carolina, Tennessee, and Virginia had more than half of their representatives opposing the freeze. Republicans from northern and industrial states gave support to the measure while southern Democrats were nearly evenly divided on the issue.

The significance of the resolution remains a debated issue among members of Congress and the public. Supporters believe it is a first step toward calls for a reduction in the arms build-up on the part of the two major powers. Opponents saw the freeze measure as a call for weakening America's defense posture and as an impediment to maintaining nuclear superiority. These opponents, including the President of the United States, felt that support for a nuclear freeze would hamper SALT (Strategic Arms Limitations Talks) discussions with the Soviets in that the U.S. would be perceived as being weak. Apart from diplomatic and political interpretations, support for the freeze reflects the public's increased concern over the likelihood of nuclear conflict between the two major superpowers and the effects such a conflict would have on the planet's populations and environments.

The Global Risk Mosaic: Depo-Provera

Depo-Provera is the trade name of a synthetic female hormone, manufactured by the Upjohn Corporation, which may be used as an ovulation-suppressing contraceptive. Probably its most attractive feature is that one injection is good for three months. Except for certain experimental uses, the Food and Drug Administration has withheld approval for the marketing of Depo-Provera in the United States. Nevertheless, a worldwide market exists for the drug which is licensed for use as a contraceptive in almost 80 nations, ranging from Europe to the Third World. These nations are listed in Table 7. Note that the table does not include such nations as Canada, Brazil, India, Japan, and Australia. In England and Ireland it is registered for only short-term anti-fertility purposes. Whether the widespread use of Depo-Provera may be interpreted as one part of the international risk mosaic is open to debate (Rosenfield et al. 1983).

Although the manufacturer believes Depo-Provera to be less harmful than estrogen as a contraceptive, the drug is not without its negative side-effects, including irregular menses and spotting, delayed return to fertility, weight gain, elevated blood glucose levels, mental depression, and hair loss. More serious allegations have also been levelled against its use and it is these allegations which have delayed its approval. Depo-Provera has been linked to cancer in animals, particularly cancer of the breast, endometrium, and cervix. It has been alleged to have a teratogenic potential as well, but probably poses no more threat in this area than do other hormonal contraceptives (Rosenfield 1983). There also exists the possibility of transfer through mother's milk to nursing infants. The company itself acknowledges potential harmful side-effects but does not list them on a warning label in its international marketing campaign.

The use and marketing of Depo-Provera on a global scale reflects a variety of decisions by individual governments. The many nations which have approved its sale indicate the success which pharmaceutical companies have had in lobbying for its approval, especially in developing countries. The nations which have denied approval, on the other hand, indicate the political clout of monitoring and regulatory agencies and their determination to protect their citizens (Tausend 1983). Some of the organizations supporting the sale and distribution of Depo-Provera are Upjohn and other phar-

TABLE 7   THE GLOBAL MARKET FOR DEPO-PROVERA

Country	Date Registered	Country	Date Registered
The Americas		**Europe**	
Antigua	1973	Belgium	1968
Barbados	1970	Cyprus	1971
Bermuda	1976	Denmark	1972
Bolivia	1969	France	1980
Colombia	1972	West Germany	1969
Costa Rica	1973	Iceland	1970
Curacao	1968	Luxembourg	1972
Dominican Republic	1970	Netherlands	1973
Ecuador	1968	Norway	1976
El Salvador	1973	Portugal	1969
Guatemala	1972	Spain	1968
Guyana	1973	Sweden	1981
Haiti	1967	Switzerland	1973
Honduras	1974	Yugoslavia	1970
Jamaica	1968		
Mexico	1967	**Africa**	
Nicaragua	1973	Cameroon	1970
Panama	1968	Egypt	1983
Peru	1978	Ethiopia	1973
Surinam	1969	Ghana	1974
Trinidad	1970	Kenya	1972
		Liberia	1973
Asia and the Pacific		Libya	1966
Bahrain	1967	Madagascar	1974
Bangladesh	1979	Malawi	1977
Burma	1969	Morocco	1975
Hong Kong	1978	Nigeria	1970
Indonesia	1976	Reunion	1974
Iraq	1970	Rwanda	1967
Kuwait	1973	Sierra Leone	1978
Lebanon	1968	South Africa	1973
Malaysia/Singapore	1968	Sudan	1974
Muscat and Oman	1979	Tanzania	1973
New Zealand	1969	Uganda	1974
Pakistan	1976	Zaire	1972
Philippines	1972	Zambia	1973
Qatar	1974	Zimbabwe	1972
Saudi Arabia	1968		
Sri Lanka	1969		
Syria	1963		
Thailand	1970		
United Arab Emirates	1980		

Source: Galligan 1983.

maceutical companies, the World Health Organization, the World Bank, the International Monetary Fund, and the International Planned Parenthood Federation. Included among those opposing its sale are Ralph Nader's Health Research Group, the National Women's Health Network, the National Organization of Women, the Congressional Black Caucus, and the United Presbyterian Church. At issue is whether the risks of encouraging widespread use of the drug are sufficiently high to prohibit it from use. A corollary issue is whether the drug may be banned for use in the United States, yet approved for distribution abroad through the U.S. Agency for International Development. Already it has been estimated that the product has been used by some ten million women, and Upjohn maintains that in such nations as Thailand, where 86,000 women have received the drug since 1965, there has been no increase in endometrial cancer. Another corollary issue is whether certain populations within the United States should be subjected to experimental applications of Depo-Provera. Upjohn, for instance, tested the drug on a number of college campuses, including the University of Washington and the University of Wisconsin-Madison, in the late 1960s and early 1970s. Testing is also going on in urban family planning clinics, mental hospitals, and retardation centers (Tausend 1983). At the Maryland State Penitentiary, it is being administered as a method of chemical castration to male inmates in an effort to gain control over those with unusual sexual appetites.

The National Risk Mosaic

Scattered throughout the United States are an estimated 32,000 to 50,000 hazardous waste dump sites, of which 1,000 to 2,000 pose an immediate threat to public health (Valoric 1981; Anderson and Greenberg 1982). Perhaps the best known of these dumps is the Hooker Chemical Company's abandoned Love Canal site in Niagara Falls, New York. Although the problem of uncontained toxic chemicals has allegedly been recognized by scientists for years, Love Canal drew nationwide public attention to the problem of how to handle waste materials which are toxic, corrosive, ignitable, or chemically reactive so that they do not leak into the environment and threaten public health. Love Canal has reshaped our knowledge of the national risk mosaic and has brought public perceptions into line with the reality of the threat, a reality which went unrecognized for decades.

In 1942, Hooker Chemical began dumping industrial wastes into an abandoned canal that was to have become the focal point of a massive industrial park. After the company purchased the canal in 1946, it became the disposal site for over 43 million pounds of such wastes as benzene hexachloride, chlorobenzenes, and trichlorphenol, all of which are highly toxic and carcinogenic (Epstein et al. 1982:92). Dumping stopped in 1952 and the company deeded the land to the local municipality on the condition it would remain undisturbed. Subsequently, the municipality permitted the local school board to build an elementary school on the site, construction of which violated the supposedly safe containment of the chemicals. Two hundred thirty-nine homes were also built there. It was not long before inhabitants began to report noxious odors, liquid sludge accumulating in basements and backyards, and barren land where nothing would grow. It took them longer to recognize the larger threats to public health which included abnormally high rates of illness, miscarriages and birth defects, congenital heart defects and incomplete skull closures, numerous crib deaths, skin rashes, bronchitis, liver and kidney problems, and, according to a federal study, chromosome damage.

After several years of confrontation between Love Canal homeowners and governmental entities at local, state, and federal levels, state authorities began the evacuation of homes in the "inner ring" closest to the canal itself. Later the evacuation advisory was extended to an "outer ring," resulting in more than 600 homes being purchased by government agencies between 1978 and 1980. Houses within the inner ring were demolished and the area was sealed off with a fence. In 1982, the Environmental Protection Agency certified the outer ring, beginning at the fence, safe for human habitation, a declaration that has rekindled the debate over the reality of risks to public health in the area. Monitoring studies that have been conducted using captured field mice in the Love Canal area over a four year time span, however, have cast doubt on the advisability of re-opening the neighborhood to settlement. John Christian (1983) and colleagues at the State University of New York found toxic chemicals lodged within the body tissues of field mice trapped near Love Canal. They also found that life expectancy among field mice, a relatively sedentary species, increased with distance from the canal. Mice living nearest the fence were found to be dying prematurely when compared with a control group farther away and with the known life expectancy of the species. Coincident with this relationship, the scientists found that the field mice were exceedingly hard to find close to the canal site even though the natural conditions there provided an ideal habitat. Moreover, their results suggested that male mice may experience delayed maturation and females reproductive problems in the areas closest to the dump.

While Love Canal may have been "our country's baptism into the problem of toxic chemical wastes" (Weinberg 1983:12), numerous places have become household words since that time because of their toxic waste problems. Woburn, Massachusetts, was the site of the state's largest chemical dump and today exhibits the highest cancer rate among Massachusetts' communities with more than 20,000 people; the childhood leukemia rate is more than double what would be expected. Hopewell, Virginia, was the site of manufacture for Kepone, a deadly pesticide, which debilitated the reproductive and central nervous systems of workers at the plant and which was discharged into the environment resulting in the closing of the James River and lower Chesapeake Bay to fishing. Montague, Michigan, served as the disposal site for waste brine, asbestos, deadly pesticides, and dioxin resulting in the complete contamination of ground water reserves and a nearby lake. Indeed, the greatest threat posed by the thousands of chemical waste dumps may be the contamination of local and regional water supplies. Leachate from improperly designed landfills may enter the groundwater and runoff may contaminate surface water. The problem of groundwater contamination by synthetic organic chemicals is widespread, but no comprehensive data bank on its geographic extent or public health impacts exists (Burmaster and Harris 1982).

More recently, Times Beach, Missouri, was found to be so polluted with dioxin that the entire town was evacuated and bought out by the federal government for $33 million. In Times Beach and other communities across the state, dioxin-contaminated waste oil was spread on roads and around stables to control dust in 1971. Not long after Times Beach, the federal government decided to buy out another Missouri community, Imperial, for the same reason. Perhaps a hundred other sites in Missouri are also contaminated by dioxin, a family of chemicals often described as "one of the most powerful carcinogens known to man." In laboratory animals, even minute concentrations of dioxin have been found to cause cancer, birth defects, miscarriages, disease of the nervous system, kidneys, and liver, and even death. Dioxin was one of the chemicals found in the neighborhoods around Love Canal and most recently in the

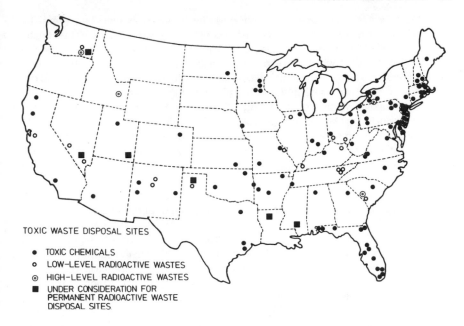

TOXIC WASTE DISPOSAL SITES

- • TOXIC CHEMICALS
- ○ LOW-LEVEL RADIOACTIVE WASTES
- ◉ HIGH-LEVEL RADIOACTIVE WASTES
- ■ UNDER CONSIDERATION FOR
  PERMANENT RADIOACTIVE WASTE
  DISPOSAL SITES

FIGURE 18   Toxic Waste Disposal Sites (Based in part on
Council on Environmental Quality 1982:202)

New Jersey communities of Newark and Clifton where it was produced on industrial sites as a by-product of chemical manufacture.

Figure 18 displays the locations of the first 114 toxic chemical waste dumps listed by the Environmental Protection Agency as posing imminent threats to public health, particularly through the contamination of ground and surface water. The EPA list has since been expanded to 546 sites. The original list was compiled from sites submitted by the individual states, a method of compilation that calls into question the assertion that these are the most dangerous dumps in the nation. New Jersey and Delaware, major chemical producing states, clearly appear to be among the most threatened states. Louisiana, however, with one of the nation's largest chemical industries, failed to submit any candidates for the EPA's list and thus is shown to be absent of dump sites on the map. Nevertheless, the release of this and subsequent lists contributes to our knowledge of the real risks presented by toxic waste disposal and alters the configuration of the perceived contoured risk surface of the United States.

Endowed with $1.6 billion and given an initial lifespan of five years, the Comprehensive Environmental Response, Compensation, and Liability Act of 1980 was designed to cover the cost of cleaning up threatening toxic waste dumps created by careless toxic chemical disposal methods over the past four score years. Identifying the locations of toxic dumps was the first step in implementing this legislation, popularly known as "Superfund." Optimal allocation of monies under Superfund can take place only if the full range of dump sites is known; otherwise, the public sector runs the risk of overlooking serious threats to public health in favor of cleaning up known but less dangerous dumps. Given the imperfect state of knowledge about the universe of places where toxic wastes have been dumped, often by "midnight haulers" or "gypsy truckers," one wonders whether the Love Canals of the next two decades will be limited to the sites on EPA's official roster. One may also question how far Superfund will go in cleaning up the sites which have been identified. So far only six dumps have been cleaned up and it has been estimated that the initial funding will cover the costs of only 170 sites. Total clean-up costs have been put as high as $40 billion.

Toxic wastes which are now being generated fall under the authority of another piece of federal legislation, the Resource Conservation and Recovery Act of 1976, which was designed to prevent substandard chemical dumps from coming into being. The RCRA assigns "cradle to grave" responsibility for chemicals that are ignitable, corrosive, reactive, or toxic. A manifest system is used to assure that waste chemicals are disposed of at approved sites.

Hazardous materials incidents present another nationwide set of technological hazards. According to a report by the U.S. Environmental Protection Agency (1980), more than 3,000 hazardous materials incidents occurred during the two-year period ending in 1979. Materials involved ranged from spills of alcohol to highly toxic polychlorinated biphenyls (PCBs). PCBs, used since the 1930s in the chemical and electric industries, were among the chemicals found seeping into basements in the Love Canal neighborhood. They may take 100 years to break down, yet are so highly toxic that one per million creates major concerns about the environment, public health, and animal husbandry industries (Brown 1981:17,278). Causes of the incidents listed in the report were linked to leaking drums, ruptured pipelines and hoses, manufacturing discharges, motor vehicle and barge accidents, explosions, transformer leaks, and illegal dumping. Amounts ranged from less than one pint to over one million gallons. As the map in Figure 19 shows, the freqency of these EPA-reported incidents is closely correlated with the distribution of population and industrial capacity. The fact that these incidents occurred most frequently in the most heavily populated states enhances their hazard potential.

Since World War II, the United States has turned out 10.2 million cubic feet of high-level radioactive reprocessing wastes; 85.8 million cubic feet of low-level wastes from commercial nuclear reactors, hospitals, research institutions, and defense facilities: 122,000 cubic feet of spent fuel assemblies from commercial nuclear power plants; and almost 3 billion cubic feet of uranium mill tailings at 19 active and 25 inactive sites (Shapiro 1981:8-12). High-level wastes are currently stored at four temporary locations; low-level wastes are buried at 22 sites (Council on Environmental Quality 1982:202). By 1990, the federal Department of Energy will select one of six possible sites as the nation's first permanent high-level waste burial site. These temporary waste disposal sites and options for a permanent one are displayed in Figure 18. The states themselves have been given the responsibility for deciding where to permanently dispose of low-level radioactive wastes. Eleven northeastern states are currently trying to jointly decide where to impound their low-level wastes, but many state legislatures have not approved the compact because they fear their state will be selected as the site of the mammoth regional facility. Outside the United States, disposal problems also lie ahead for the United Kingdom, West Germany, France, and Japan. The 1980s are sure to be a decade of locational conflicts over radioactive waste disposal siting. Even the oceans will become part of the controversy as the U.S. Navy seeks sites at which to dispose of more than 100 nuclear-powered submarines that will be decommissioned over the next thirty years. The alternative, albeit hundreds of millions of dollars more expensive, would be to haul contaminated portions of these vessels to land-based disposal sites.

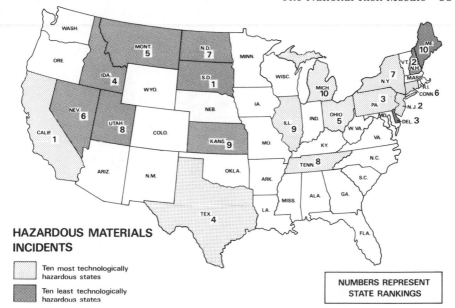

HAZARDOUS MATERIALS
INCIDENTS

☐ Ten most technologically
hazardous states

▨ Ten least technologically
hazardous states

NUMBERS REPRESENT
STATE RANKINGS

FIGURE 19   Hazardous Materials Incidents, 1977-1979 (Based on
U.S. Environmental Protection Agency 1980)

The risk surfaces generated by these high-level and low-level radioactive waste
dumps are not limited to the areal hazard zones surrounding the sites themselves. In
addition, a set of linear hazard zones will be engendered by the transportation of
radioactive wastes from their place of production to their place of disposal. Many of the
linear hazard zones which are currently part of the national risk mosaic are presented in
Figure 20, which shows the highways used to transport uranium to enrich and fuel
fabrication facilities, fuel to nuclear reactor sites, and spent fuel to storage facilities. The
risks of radioactive waste transportation have yet to be adequately assessed, however.
In Virginia, testimony to the existence of some risk is the federal and state government's decision to route radioactive waste through more sparsely settled parts of the
state rather than using Interstate highways 95 and 395 which traverse heavily populated corridors. Furthermore, only one port on the east coast, Portsmouth, Virginia, will
allow radioactive wastes from European research reactors to pass through its facilities.
This practice began with Eisenhower's Atoms for Peace Program and is scheduled to
be expanded to include spent fuel from commercial reactors located in foreign nations
in an effort to promote non-proliferation (Shapiro 1981:76). Through the streets of
Portsmouth and nearby Suffolk head the trucks carrying radioactive wastes to South
Carolina. As this information has been brought to light by a local environmental group,
the perceptual risk map of Tidewater has been restructured and one wonders how the
heightened perception of risk compares with the reality of the risk surface.

Another low probability/high consequence technological hazard to which the
nation is exposed is the threat of nuclear holocaust. The map in Figure 21 shows the
radioactive fallout regions that might result from an attack on the U.S. delivering 6,600

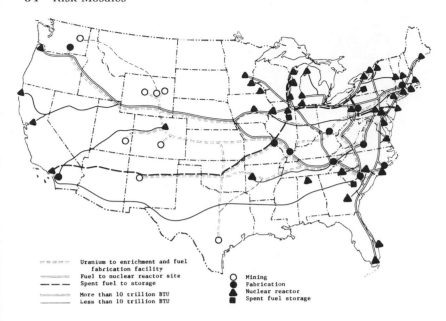

FIGURE 20   Linear Hazard Zones Resulting from the Transportation of Nuclear Materials (Congressional Research Service 1977)

megatons, comprising 113 surface blasts of 20 megatons each on urban and industrial targets, a total of 3190 megaton surface blasts on military targets, and 1150 megaton air bursts which would produce practically no fallout (Kearney 1982:23). The map displays distinct plumes away from the sites of possible detonations. The highest risk areas would be ICBM sites near Great Falls, Montana, and Tucson, Arizona; Strategic Air Command and Naval submarine bases; major ports and airports; and munitions factories. In addition, cities with steel mills, oil and chemical refineries, would also be high-risk targets. The eastern half of the country, along with California, are likely to experience heavy radiation doses following multiple surface blasts. Sparsely populated areas of the west appear safest, but as Kearney (1982:25) points out, as the number of warheads aimed at the U.S. increases, the areas that "could be free of fallout probably would be less extensive."

The speculative fallout patterns on the map may be used as rough guides to "help improve chances of evacuating a probable blast area or a very heavy fallout area and going to a less dangerous area" (Kearney 1982:23). Ultimately, the United States hopes to have "relocation plans" for 80 percent of the American people. They would be directed in the event of an anticipated attack to small towns and rural areas at least 20 miles away from the target areas. The money that will be pumped into evacuation planning during the next several years has been promoted as a two-pronged risk management strategy. First, evacuation is designed to protect the American people during a nuclear attack. Second, having evacuation plans which can be quickly and efficiently put into action may reduce the risk of nuclear war itself by turning cities into

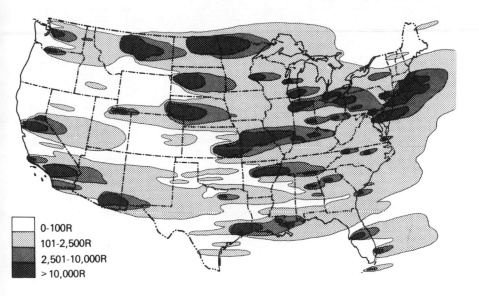

	0-100R
	101-2,500R
	2,501-10,000R
	> 10,000R

FIGURE 21   Hypothetical Radioactive Fallout Regions (Redrafted from Kearney 1982:24; originally prepared
by Oak Ridge Geographics)

empty shells hardly worth bombing. If nothing else, such rhetoric and the elaborate preparations for evacuating America's metropolitan areas "reinforces the sense that we remain in the same political, organization, and moral climate that evolved through the two World Wars" (Hewitt 1983:279).

Preparations for nuclear war have also seen 801 nuclear tests take place throughout the world between 1945, when the world's first nuclear bomb was detonated at Alamgordo, New Mexico, and June 1975, all of which are listed by Carter and Moghissi (1977). In the United States most tests have taken place at the Nevada Test Site, the Pacific Proving Ground, and several other islands of the Pacific. Although isolated, sparsely populated areas were selected for these tests, their negative side effects have been imposed on unsuspecting populations. The risks of above-ground testing were not well understood; risk management strategies were therefore not commensurate to the task. Sternglass (1969) found considerably elevated levels of infant mortality downwind from the Alamgordo site and from Arctic test sites (Bunge 1973). These elevated rates of infant mortality are displayed in Figure 22. Birth defects, miscarriages, and cancer have also been found to characterize populations in the vicinities of nuclear test sites. Lyon and colleagues (1979) studied the effects of radiation on the childhood population in the wake of 26 nuclear tests which took place in the Nevada desert between 1951 and 1958. They discovered a higher incidence of childhood leukemia among those who resided in high-fallout counties while they were under the age of 15.

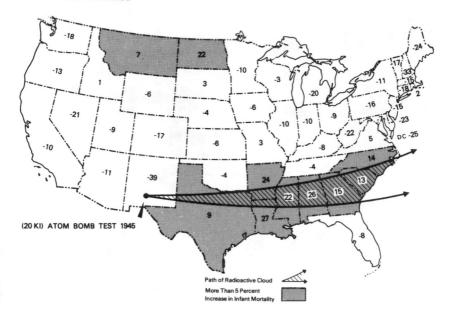

(20 KI) ATOM BOMB TEST 1945

Path of Radioactive Cloud
More Than 5 Percent
Increase in Infant Mortality

FIGURE 22   Deviation of Infant Mortality from Expected, 1950
(Redrafted from Bunge 1973; based on Sternglass
1969; courtesy of *Medical Tribune)*

The State and Local Risk Mosaic

A case study approach will demonstrate the utility of part of the technological hazards typology presented in Chapter 1 and will illustrate the magnitude of the problem at the regional scale in California. Data gathered by the Environmental Protection Agency (1980) and displayed in Figure 18 reveals that California reported more hazardous waste incidents to the EPA between 1977 and 1979 than any other state, accounting for 303 incidents or about 10 percent of the total. The EPA report includes information on the type, date, location, and material spilled in each incident, as well as the environmental medium affected. For the purposes of this study, the incidents reported in California were classified according to whether they occurred in the production or transport and transmission of hazardous materials. When the cause of the incident was not reported, it was classified as "unknown." The locational information made it possible to assign each incident to a specific metropolitan or non-metropolitan area but the following analysis is restricted to the 264 incidents which occurred in the state's Standard Metropolitan Statistical Areas (SMSAs).

Nearly half (46 percent) of the 264 incidents involved the transport and transmission of hazardous materials, resulting from such accidents as train derailments and truck accidents. Approximately four out of every ten incidents occurred in either the production of goods that involved the use of hazardous materials or in the actual production of hazardous materials. The remaining 16 percent fell into the "unknown" category. Figure 23 shows that at least one incident involving hazardous materials

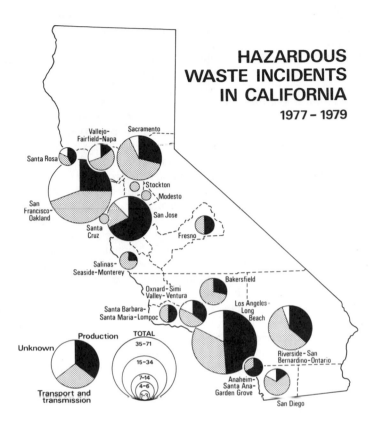

FIGURE 23   Hazardous Waste Incidents in California, 1977-1979
(Based on U.S. Environmental Protection Agency 1980)

occurred in each of California's 17 SMSAs between 1977 and 1979. The spatial distribution of incidents was strongly correlated ($r = +.83$) with the population distribution. That is, the largest number of incidents tended to occur in the major population centers, while fewer incidents were recorded in the smaller SMSAs. Over half of the 264 incidents occurred in the San Francisco-Oakland and the Los Angeles-Long Beach SMSAs, where nearly two-thirds of the state's population resided in 1978. In contrast, fewer than three incidents were reported in the much smaller Salinas-Seaside-Monterey and Santa Rosa SMSAs.

The distribution of incidents was also strongly associated ($r = +.73$) with the distribution of industrial establishments. This is clearly evident upon examining the distribution by type in Figure 23. Over half of the incidents involving the actual production of hazardous materials or of goods in which hazardous materials are used in the manufacturing process, occurred in the much more industrialized SMSAs of southern California (especially in the Los Angeles-Long Beach metropolitan area) than in the SMSAs of the northern part of the state (north of the Salinas-Monterey SMSA). More than 70 percent of the state's industrial activities are concentrated in southern California. By contrast, most of the transportation-related incidents occurred in the less industrialized SMSAs of northern California. The major exception to this general pattern is the San Jose SMSA of northern California where over two-thirds (68 percent) of the incidents were production-related.

In short, incidents involving hazardous materials occurred in metropolitan areas of California at an average rate of 11 per month between 1977 and 1979. Whereas in northern California the people residing along major transport arteries were more likely to be exposed to a hazardous materials incident, southern California households living near primary and secondary production sites were most likely to be subjected to such hazards. To alleviate the risks to which these people are exposed several steps need to be taken. First, improved operation, maintenance, and monitoring of equipment used in the transport and transmission of hazardous materials would go a long way toward eliminating these kinds of hazards. Second, stronger enforcement of safety standards in and around workplaces would minimize hazardous materials incidents in the primary or secondary phase of the production of consumer goods. Third, stricter guidelines pertaining to the reporting of incidents involving hazardous materials incidents to the EPA are needed since the "cause of incident" in approximately 15 percent of the cases reported between 1977 and 1979 was unknown. Stiff penalties for failure to provide, within a specified period of time, complete information on the place, date, and type of incident, material spilled and volume, and the environmental medium affected would help solve the problem of incomplete and often haphazard reporting.

Some of the hazards which plague the local environment are generated at the global or regional scale, while others arise within the locality itself. Indigenous urban hazards include such occurrences as automobile accidents, concentrated pollutants, and noise. Combining these with natural and other technological hazards that affect the local setting makes it possible to define high-risk and low-risk areas of a city. Hewitt and Burton (1971) proposed an ecological research paradigm which stresses the need to define the total "hazardousness of a place" by enumerating, mapping, and defining the relationships and management possibilities for all hazards to which a place is exposed. Foster (1980) suggested the spatial manifestations of such an effort would be a map showing zones of high and low risk — "hazard microzoning." Such total risk maps could then be used as risk-management tools in land use and activity planning. Some areas, such as those around waste dumps, might remain permanently unsettled, while other areas, such as flood plains, might be permitted to undergo low-intensity development.

Several components of the local risk mosaic in the Los Angeles area include sites of hazardous waste incidents and waste dumps (Figure 24; U.S. Environmental Protection Agency 1980). The waste incidents have been classified according to the typology proposed in Chapter 1. Some incidents accompany the production process and others result from the transportation of hazardous materials. The pattern shows clearly the hazards arising from toxic materials are heavily concentrated in certain parts of the city. Several research questions therefore present themselves: Are local residents aware of the incidents and dump sites? How do they perceive the risks? What have been the epidemiological consequences of the hazard? What social groups have been most severely affected by residing in proximity to these sites? How has the pattern of incidents and dump sites affected property values and development?

One of the worst dumps in California is located 20 miles east of Los Angeles in Fullerton, only a few miles from Disneyland. It is the McColl dump, an abandoned disposal site which is the resting place for 50-75,000 tons of chemical sludge from the manufacture of aviation fuel during World War II. The sludge is charged with sulfuric acid and heavy metals. In the newly settled residential neighborhoods nearby, back-

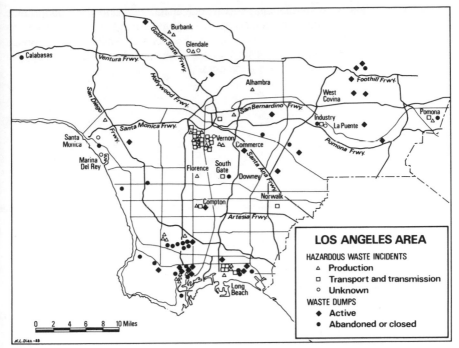

FIGURE 24   Hazardous Waste in the Los Angeles Area (Based in part on U.S. Environmental Protection Agency 1980)

yards have flooded with bubbling ooze during heavy rains, the air is permeated with noxious odors and as many as 50 chemical substances, and residents have been found to exhibit excessive respiratory, eye, and stomach problems. Bottled water and air filtration systems for nearby homes have become part of everyday life as property values have declined in the immediate vicinity. According to a social survey conducted by one of the authors, the predominantly white, middle- and upper-income residents of the area registered high levels of opposition to the McColl dump. Opposition was uniformly high within a six block radius of the site. It is believed that this is the zone in which odors emanating from the dump are most apparent. More than 80 percent of the residents felt that the dump was having a negative impact on their neighborhood. Those residents closest to the site also felt that the dump was directly responsible for current health problems experienced by members of their families, particularly allergies, headaches, nausea, eye irritation, and respiratory difficulties.

In every community there are numerous hazards of technological origin to which the public is exposed. Some threaten physical health and others mental heath, some degrade natural environmental systems and others the quality of the residential environment, some have ready solutions that can be implemented by individual citizens while others require the power and resources of higher levels of government, and virtually all make it just that much more difficult for communities to progress up the needs hierarchy to the level of community-actualization. This chapter has presented only a few examples of hazards at the global, national, state, and local scales. Whereas the previous three chapters have provided the necessary background and some conceptual models for understanding technological hazards and their geographic dimensions, this chapter has provided some factual background on such hazards as nuclear war, Depo-Provera, toxic chemical dump sites, and radioactive waste disposal. The next chapter takes a look at a specific problem in the hazards arena, the problem of evacuation from nuclear power plant accidents.

5

Evacuation: A Response to Technological Disasters

In 1982, 20,000 inhabitants of Taft, Louisiana, evacuated their homes after a chemical tank exploded; 250 residents of Los Banos, California, fled after a truck carrying nine guided missiles overturned; and a year earlier 2,600 evacuees left their homes in Bridgman, Michigan, after fluorosulfuric acid fumes escaped into the atmosphere from a derailed train. The accident at Three Mile Island precipitated the evacuation of more than 144,000 people, and a train derailment in Mississauga, Ontario, resulted in the evacuation of an estimated quarter of a million. Unlike the previous instances, the residents of Love Canal, New York, and Times Beach, Missouri, have been permanently evacuated from their homes as a result of technological hazards.

Hans and Sell (1974), in what remains the only comprehensive study of evacuation, reported that an average of almost 90,000 persons per year were involved in an evacuation during the 1960-1973 period. Included in their review and evaluation were both natural and technological events; however, most studies of the evacuation process have been specific to natural disasters (Rayner 1953; Moore et al. 1963; Drabek and Boggs 1968; Drabek 1969; Drabek and Stephenson 1971; Mileti and Beck 1975; Perry 1979; Perry et al. 1981; Green et al. 1981).

Perry (1979:440,446) called evacuation "an important management tool for minimizing the catastrophic consequences of natural disasters in an orderly way," but warned that "it is important to build emergency planning around people's known reaction patterns." Unfortunately, planning for all types of hazardous events has tended to proceed on the basis of what is known about human responses to natural disasters. When planning for technological hazards, a fundamental question arises: How pertinent is behavior during natural disasters to technology-induced emergencies?

In his comprehensive review of natural disaster studies, Perry (1979) concluded that the following variables were critical in an individual's decision to evacuate: community and family context; prior experience with disasters; warning source, frequency, and context; presence of an adaptive plan; perception of the threat as real; and the level of perceived personal risk. From these variables he formulated a causal model of evacuation decision-making. In a later work (Perry et al. 1981) this model was employed as the basis for a path analysis to determine how well the variables explained the decision to evacuate in four flood-stricken communities in the western United States. Perceived threat and personal risk were found to be the most important of the explanatory variables considered. Not incorporated into the model, however, were spatial and temporal variables such as the impact of distance from the disaster agent of risk perception and the space-searching behavior of evacuees.

Evacuation, like migration, may be viewed as a response to environmental stress (Wolpert 1966); it is fundamentally a spatial process whereby an individual exchanges a high-stress for a low-stress location. "Few events," said Wolpert (1980:393), "seem as threatening as the need to be evacuated or displaced permanently from one's home." Within this context, evacuation planning must be based on a thorough understanding of people's perceptions, attitudes, concerns, and behaviors under stressful environmental conditions. If a comprehensive theory of disaster-response behavior is to be developed as a basis for emergency planning, the differences and likenesses between natural and technological disasters must be understood within a spatial context. Even among technological disasters there seems to be a wide gulf separating evacuations from spreading chlorine gas clouds and evacuations from nuclear power plant accidents. Since nuclear threats are among the most potentially severe of all technological hazards, they will constitute the focus of this chapter.

Some of the dimensions of evacuation planning are summarized in Table 8 which enumerates a set of selected spatial, temporal, demographic, and operational characteristics of the evacuation process. In terms of scale, evacuations may range from individual buildings, to neighborhoods, to metropolitan areas, to larger regions. The scale of the evacuation has serious implications for emergency planning in terms of complexity. The evacuation of a school or office is far less complex than the evacuation of an entire city; it is also more common. A subsidiary consideration is the dimensionality of evacuation. In general, most evacuations take place in the horizontal dimension, but within high-rise buildings the vertical dimension may be crucial. In the case of storm surges along the coast or fires it may be necessary to evacuate to the top of buildings to either weather the crisis there or be air-lifted off. Likewise, the search for high ground during flooding shows how the vertical dimension may influence evacuation decision-making.

Each of the continua presented in the diagram has important planning implications and can make a difference in the type of evacuation which is necessary. At Love Canal and Times Beach permanent evacuations were necessary; in Taft, Louisiana, the evacuation was only temporary; at Three Mile Island most evacuees returned within a week, but many who left wondered whether their communities would be rendered permanently uninhabitable while they were gone. The timing of evacuation is also critical: If there is not enough time to stage a full-scale, pre-impact exodus, a delayed evacuation may be selected since evacuees are sometimes more vulnerable to harm if the impact occurs while people are on the road and unprotected (Foster 1980:226). The characteristics of the population to which the evacuation order applies also exercise an influence on the evacuation process as does the specificity of the evacuation plan itself.

TABLE 8   DIMENSIONS OF EVACUATION

	Spatial Dimensions	
Local	SCALE	Regional
Vertical	DIMENSIONALITY	Horizontal
	Temporal Dimensions	
Temporary	DURATION	Permanent
Pre-Impact	IMPACT	Post-Impact
Immediate	TIMING	Delayed
	Demographic Dimensions	
General	COMPREHENSIVENESS	Selective
Urban	SETTLEMENT PATTERNS	Rural
Diurnal	FLUCTUATIONS	Seasonal
Homogeneous	SOCIOECONOMIC STATUS	Heterogeneous
	Operational Dimensions	
Simple	ADAPTIVE PLAN	Complex
Non	PRIOR EXPERIENCE	Extensive
Voluntary	VOLITION	Ordered
Visible	HAZARD AGENT	Invisible
Reliable	INFORMATION	Unreliable

Planning for Radiological Emergencies

The purpose of radiological emergency response planning is "to provide dose savings for a spectrum of accidents that could produce offsite doses in excess of Protective Action Guides (PAGs)" (U.S. Nuclear Regulatory Commission 1980:6). A PAG is a projected dose of radiation which would warrant some type of protective action. This action may take the form of evacuation, sheltering, the administration of a thyroid blocking agent, or some combination of the above. Sheltering would require that people remain in their homes with windows and doors closed and air intakes sealed off. It is likely to be ordered if release time is expected to be short, if there is not enough time to complete an evacuation, or if an area is beyond the evacuation zone. The oral administration of potassium iodine may also be recommended in order to saturate the thyroid gland with enough stable iodine to prevent it from taking up iodine isotopes.

Before the accident at Three Mile Island, emergency preparedness plans were required for only the 2-3 mile low population zone (LPZ) around the reactor site. The chances of a serious accident at a nuclear power plant were judged to be remote, and even if an accident were to occur, offsite consequences were judged to be improbable. Radiological planning was therefore not "in a position of high visibility within the nuclear industry or within the Federal, State, and local governments" (Federal Emergency Management Agency 1980:I-2). After TMI, however, the NRC, jointly with the Federal Emergency Management Agency, issued revised regulations (U.S. Nuclear Regulatory Commission 1980) which require all licensees to develop extensive offsite plans in cooperation with local and state governments. These plans are to be based on two exposure pathway zones around the nuclear site:

(1) *Plume Exposure Pathway Emergency Planning Zone* — Encompasses a radius out to ten miles from the plant. This is the only zone for which evacuation planning is required since the "probability of large doses [of radiation] drops off substantially at about 10 miles from the reactor" (U.S. Nuclear Regulatory Commission 1978a:I-37). The primary hazard within this zone would be from direct exposure and inhalation of radioactive materials.

(2) *Ingestion Exposure Pathway Emergency Planning Zone* — Encompasses the area between 10 and 50 miles from the plant. The State is responsible for emergency planning in this zone. The primary hazard here would be from ingesting contaminated food and water.

The areas of the United States which fall within the 50-mile ingestion exposure pathway EPZs are shown in Figure 25. These are the zones most likely to be affected by a commercial nuclear power plant accident. At highest risk are the areas which are shown to be within the EPZs of two or more nuclear plant sites. The correlation with the urban settlement pattern is striking, particularly along the northeastern seaboard, around the Great Lakes, and in the Carolina piedmont. Within 50 miles of the Indian Point plant in New York, for instance, reside 15 million people; around the Zion and Dresden, Illinois, plants reside 7 and 6.5 million respectively; and around the Shoreham plant on Long Island reside over 5 million. Indian Point and Zion also head the list of plants with the largest populations in the 10-mile planning zone. Nationwide, about 3.5 million people reside within 10 miles of nuclear reactors and 85 percent of the existing and planned nuclear generators are within 60 miles of an SMSA (U.S. Nuclear Regulatory Commission 1977). Opinion polls conducted during the 1970s showed that the public is less enthusiastic about nuclear reactors near their home communities than they are about nuclear power in general. It is therefore not surprising that easterners have shown themselves to be more concerned about safety than the citizens of other regions (Farhar *et al.* 1980:150-151).

Figure 25 also shows that the areas at highest risk are located in parts of the United States which are beset by high degrees of geopolitical fragmentation. The Middle Atlantic and Great Lakes states are perhaps more broken up into numerous local jurisdictions than any other area of the country. Compounding the problems of coordination and information flow presented by the many local governments are the state and federal agencies involved in emergency response. Over 150 federal, state and local agencies were involved in the TMI emergency, for instance (Chenault *et al.* 1979:viii).

Following up on the recommendation of the President's Commission on the Accident at Three Mile Island (1979), emergency response planning is not to be based on only one projected chain of events; it is to be flexible enough to be adapted to a spectrum of emergencies. In addition, all plans must include provisions for three populations: permanent residents, transients (e.g., tourists), and the institutionalized population. Response plans for individual reactor sites are to be formulated and evaluated upon the quality of their procedures for (1) assigning responsibility for emergency management to particular agencies; (2) defining onsite emergency organization; (3) identifying emergency response support and resources; (4) adhering to a standard classification of emergencies — unusual event, alert, site emergency, general emergency; (5) establishing procedures for notifying all agencies and personnel responsible for emergency response; (6) providing for prompt communications among principal emergency personnel and the public; (7) making information available to the

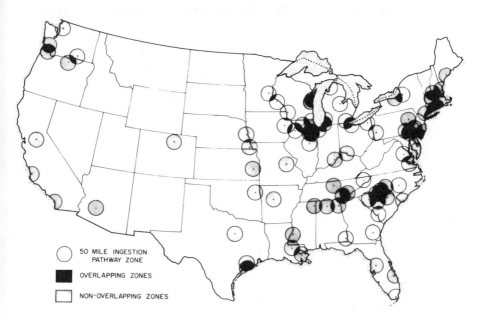

FIGURE 25   Ingestion Exposure Pathway Zones for U.S. Nuclear
Reactors

public on a periodic basis; (8) making sure that emergency facilities and equipment are
adequate; (9) monitoring off-site consequences of an emergency; (10) developing a
range of protective actions for those within the plume exposure pathway; (11) controlling the level of exposure to which onsite workers are exposed; (12) making
arrangements for medical services for contaminated individuals; (13) conducting
periodic exercises and drills to test response capabilities; (14) training emergency
response personnel; and (15) assigning responsibility for plan development and review
(U.S. Nuclear Regulatory Commission 1980).

The Evacuation Shadow Phenomenon

The first general emergency at a commercial nuclear reactor occurred in 1979 at
Three Mile Island near Harrisburg, Pennsylvania. Two days after the initial malfunction,
in response to a large release of radioactive steam, the Governor of Pennsylvania
issued an evacuation advisory for pregnant women and pre-school children living
within 5 miles of the plant, and a sheltering advisory for everyone else within ten miles.
Although Wolpert (1977) had earlier considered the theoretical problems of evacuation
from a nuclear accident, the conditions at TMI provided social scientists with their first
opportunity to study overt evacuation behavior in response to a nuclear technological
disaster and to compare with behavior during natural disasters. One of the most

significant impacts of the TMI accident was the materialization of an extensive "evacuation shadow" extending far beyond the 5-mile evacuation zone or even the 10-mile sheltering zone. The evacuation shadow phenomenon has been defined as "the tendency of an official evacuation advisory to cause departure from a much larger area than was originally intended" (Zeigler et al. 1981:7). If only those people who had been advised to evacuate had left, the number of evacuees would have been limited to an estimated 3400 women and children living within 5 miles of the reactor (Goldhaber and Lehman 1983). Instead, an estimated 144,000 individuals (39 per cent of the population) within 15 miles of TMI (Flynn 1979), and thousands more beyond 15 miles, left their homes. Our survey revealed an evacuation rate of 9 per cent as far as 25 miles away from the plant (Zeigler et al. 1981:7). We are unaware of any other disaster in which the suggested evacuation of so few resulted in the spontaneous evacuation of so many.

In addition, the median evacuation distance recorded for TMI evacuees was longer than for any previous disaster. The longest median distance cited by Hans and Sell (1974) was 80 miles in response to Hurricane Carla. In response to the accident at TMI, the median evacuation distance was recorded at 85 miles in one study (Zeigler et al. 1981) and at 100 miles in another (Flynn 1979). The mean evacuation distance recorded in a Rutgers University study was 112 miles (Cutter and Barnes 1982). Both the magnitude of evacuation and the distance evacuees travelled seem to betray a perception of nuclear power's threat that is different from general perceptions of risk from natural and other technological hazards.

Later in 1979 another type of technological disaster in Missisauga, Ontario, precipitated the evacuation of even more people than at TMI. An estimated 255,000 inhabitants left their homes in response to a train derailment causing a BLEVE (Boiling Liquid Expanding Vapor Explosion) and threatening to rupture a liquid chlorine tanker. The ordered evacuation of residents has been documented by Liverman and Wilson (1981) who found that because one-third of the evacuees did not establish themselves far enough away from the hazard agent, they had to be advised to evacuate at least one more time as the hazard zone expanded. In addition, the destinations of evacuees were largely confined to the Toronto metropolitan area and the Toronto-Hamilton corridor. The threat of an expanding cloud of chlorine gas seems to have precipitated a dramatically different behavioral response than the threat of radiation at TMI. In Ontario the evacuation shadow was limited in scale and the distances evacuees travelled were far shorter. As Liverman and Wilson (1981:365) pointed out, studies such as these may someday make it possible to develop "a model of public decision-making under threat which would identify and link critical factors, decisions, and behavior and which would provide a framework for emergency planning and research."

Since the accident at Three Mile Island, evacuation planning has become a necessary preoccupation of utility companies and local and state governments. For a new plant to be granted an operating license, emergency response plans satisfactory to FEMA, the local and state governments, and the utility must be in place. Still, most plans that have been approved have not been built on a solid behavioral foundation which takes into account the differences in response patterns between nuclear and other disasters. A report done for FEMA evaluating evacuation planning in the aftermath of TMI made the following recommendation, on which there has been little action:

*The problems of spontaneous evacuation and anticipating public be-
havior and response should be further analyzed. State and local agencies need methods for estimating the flow and extent of spontaneous
evacuation movements from the time a crisis begins (Chenault et al.
1979:x).*

An example of the difficulties presented by ad hoc planning under crisis conditions with no sensitivity to human behavior is provided by the plan drawn up by Cumberland County in the few days after the initial malfunction at Three Mile Island. No part of the county is within the 5-mile zone but the southeast corner is within 10 miles of the reactor. The County apparently conceptualized the problem of evacuating people closest to the plant as a matter of logistics rather than of human response to an emergency much different from the floods to which this area is prone. Basic features of the County plan included: (1) using schools as "staging areas" where all residents of a particular zone would have to assemble before they would be allowed access to the evacuation routes; (2) forcing evacuees to move in vehicular convoys on pre-determined routes to pre-determined host areas; and (3) barricading the five roads leading into the County from neighboring Dauphin and York Counties which are the closest counties to TMI (Chenault *et al.* 1979: 101-102). These procedures never had to be put into operation because the evacuation zone was never expanded to 10 miles, but they do illustrate the types of regulations that may evolve when human perceptions and behaviors are not factored into the emergency planning process.

As a result of the findings at Three Mile Island, New York's Suffolk County commissioned a social survey of Long Island residents as a basis for developing a plan to cope with an emergency at the Shoreham nuclear power station, 60 miles from New York City on Long Island Sound. The Shoreham plant has become one of the most controversial and, at an estimated $3.2 billion, most expensive plants in the history of nuclear power. Two of the authors participated in the research design and analysis of results from the Shoreham survey which was conducted by telephone with 2595 families in Suffolk and Nassau counties (Johnson and Zeigler 1982; Social Data Analysts 1982a).

The overriding objective of the survey was to develop a behavioral data base for emergency response planning. One of the specific objectives was to be able to predict the magnitude and areal extent of the evacuation shadow phenomenon on an island where, as a result of its peculiar configuration, limited avenues of egress, and dense population, the public is rightfully concerned about being able to evacuate. Although the survey was able to elicit only behavioral intentions, Ajzen and Fishbein (1980) have shown that a person's intended behavior is the best predictor of actual behavior as long as intention does not change before the individual has a chance to act the behavior out. Further corroborating the findings of the Shoreham survey is the fact that the intended behaviors of Long Island residents closely parallel the actual behaviors of TMI area residents in 1979 (Zeigler and Johnson 1984).

Participants in the Shoreham survey were presented with the following scenarios and asked how they would react. Each scenario was prefaced by the statement "Suppose that you and your family were at home and there was an accident at the Shoreham nuclear power station. . ."

(1) Scenario I: All people who lived within 5 miles of the plant were advised to stay indoors.

(2) Scenario II: All pregnant women and pre-school children living within 5 miles of the plant were advised to evacuate and everyone else within 10 miles of the plant was advised to remain indoors.

(3) Scenario III: Everyone living within 10 miles of the plant was advised to evacuate.

In response to each scenario, participants were asked to choose one of these responses: go about your normal business, stay inside your home, leave your home and go somewhere else, or don't know. The results of the survey are presented in Table 9.

Especially conspicuous in the table is the extent of over-reaction to limited protective action advisories. This over-reaction, called 'hypervigilance' by some (Janis 1962) and 'the counter-disaster syndrome' by others (Wallace 1956; Erikson 1982), is responsible for the evacuation shadow phenomenon as it shows up in the results of the survey (Johnson and Zeigler 1983). In response to Scenario I, in which no one was advised to evacuate, 40 percent of the population within 10 miles and 25 percent of the total population of Nassau and Suffolk Counties said they would leave. In response to Scenario II, which was modeled after TMI, only nearby pregnant women and pre-school children were advised to evacuate, yet 54 percent of all families within 10 miles and 34 percent of the total population indicated they would choose to evacuate. In response to scenario III, in which everyone living within the plume exposure pathway EPZ was advised to evacuate, 78 percent of the population within 10 miles and 50 percent of the total population said they would leave their homes in search of safer quarters. In absolute number, 215,000 families indicated they would take to the road in scenario I, 289,000 in scenario II, 430,000 in scenario III.

These figures have several implications for the emergency planning process. First, it cannot be assumed that people will follow the directions of public officials during a nuclear emergency. Plans that expect people to respond as told are likely to be less effective than they could be. Unlike the Mississauga disaster during which over 90 percent (Liverman and Wilson 1981:367) of those ordered to evacuate did leave the hazard zone, the Shoreham survey found that in Scenario III, 22 percent of the families

TABLE 9   RATES OF INTENDED EVACUATION ON LONG ISLAND[a]

Area	Scenario I	Scenario II	Scenario III
5-Mile Zone	40	57	78
6-10 Mile Zone	40	52	78
Eastern Suffolk County	22	30	46
Western Suffolk County	34	44	63
Nassau County	18	25	39
Total Population	25	34	50

Source: D. J. Zeigler and Social Data Analysts 1982a.
[a]Data in percentages.

within 10 miles would not evacuate when told to do so, yet 49 percent beyond the 10 mile zone would evacuate. Second, planning for an evacuation from only the 10-mile plume exposure pathway EPZ is underplanning for an emergency since less than one out of twenty evacuees would originate from within that zone. At Mississauga, virtually all evacuees originated within the zones for which evacuation was ordered. The danger of not taking spontaneous evacuation into consideration is that the number of families evacuating from beyond the 10-mile evacuation zone may hamper the speedy removal of the families at higher risk closer to the plant.

Role Conflict Among Emergency Personnel

Another assumption that may not apply to nuclear disasters is that emergency workers will report promptly to carry out the range of duties assigned them. Emergency planning regulations for nuclear power plants fail to consider the potential impact of multiple group membership on the behavior of designated emergency personnel in crisis situations. Even studies of non-radiological emergencies have shown that some emergency workers experience what is referred to as role conflict. As Killian (1952: 310-311) noted more than three decades ago:

> When catastrophe strikes a community, many individuals find that latent conflict between ordinarily nonconflicting group loyalties suddenly becomes apparent and that they are faced with the dilemma of making an immediate choice between various roles. . . . Indeed, only the unattached person in the community was likely to be free of such conflict.

Killian (1952:311) found that the most common conflict reported in the four disaster-stricken communities which he studied was loyalty to the family versus community obligations. Policemen, firemen, and public utility workers found themselves torn between serving a community which needed their special abilities and taking care of their loved ones who were threatened by the same disaster. In his study areas, the conflict was most often resolved in favor of looking after the health, safety, and welfare of the family.

The evidence amassed in previous disaster studies suggests that role conflict may result in either of two behavioral responses on the part of emergency workers: (1) delayed response, in which the individual reports for duty only after ascertaining (through direct or indirect contact) that family members are safe, and (2) non-response, in which the individual relocates family members from the danger zone and stays with them for the duration of the crisis.

Evidence from the Three Mile Island accident suggests that role conflict may be an especially serious problem during nuclear emergencies. Smith and Fisher (1981) reported that one of the crucial problems at local hospitals (none of which was located in the 5-mile zone) was the shortage of staff resulting from the evacuation of physicians, nurses, and technicians. Maxwell (1982:276) also found that at local hospitals "the conflicting responsibilities to family and work resulted in escalating staffing problems as the crisis continued." Some hospital personnel, primarily workers and managers, moved their families outside the danger zone and returned to work (delayed response). Others, primarily doctors, nurses, and staff with young children, left the area and stayed away until the crisis was over (non-response). At one of the local hospitals only six of the 70 physicians scheduled for week-end emergency duty reportedly showed up for work. In case of another radiological emergency, "administrators can expect significant absence from staff members who have family responsibilities and should anticipate a shortage of physicians as well" (Maxwell 1982:278). According to Chenault and others (1979:139) in their assessment of the TMI evacuation, the spontaneous evacuation of hospital and nursing home staff may lead administrators to evacuate these institutions early out of fear that their staffing levels will be inadequate when the time comes. This action could trigger larger movements by the general public.

Hospital personnel were not the only people who experienced role conflict during the TMI crisis. Problems of multiple group loyalties reportedly also occurred among TMI nuclear power plant personnel. In a comparative study conducted six months after the accident of the perceptions, attitudes, feelings, and behavioral responses of super-visory and non-supervisory personnel at TMI and the Peach Bottom plant (40 miles away), Kasl and colleagues (1981:478) noted a greater degree of role conflict among TMI workers as they tried to decide how to be in two different places at the same time and how to arrive at a decision with their spouses regarding the family's response to the accident. For non-supervisory personnel, the overwhelming majority (about 90 per-cent) of the conflict involved family role versus work role. These researchers also discovered that approximately 4 percent of the TMI supervisory personnel and eleven percent of the non-supervisory staff evacuated during the reactor crisis. Most of those left the area within five miles of the plant.

Additional evidence of the extent to which designated emergency personnel may experience role conflict comes from surveys of bus drivers and firemen in the vicinity of New York's Shoreham nuclear power plant and of public school teachers in the vicinity of California's Diablo Canyon reactor. In these surveys respondents were asked: "As a designated emergency worker, what do you think you would do *first* if there was a reactor accident in which everyone within ten miles of the plant was advised to evacuate?" The response options were (1) assist with the evacuation; (2) make sure family was safe; (3) leave the area immediately; and (4) do something else.

As Table 10 shows, more than two-thirds of the school bus drivers and firemen indicated that loyalty to the family would take precedence over their duties as emer-gency workers. In contrast, less than one-fourth of San Luis Obispo County's school teachers said they would check on family members first; most said they would first help evacuate school children from the designated hazard zone. A significant proportion of this later group qualified their responses, however, by stating that participation would be (1) contingent upon being able to contact family members by telephone; (2) re-stricted to the evacuation of their class only; or (3) limited to a specified length of time. Several of these caveats do not augur well for evacuation planning. During the Three Mile Island accident, two million calls were attempted on a system designed to handle only half that number (Chenault et al. 1979; Maxwell 1982). Under such conditions, teachers who cannot contact other family members may not remain with their classes. In addition, teachers who are willing to evacuate only their classes or participate for only a limited period of time could severely hamper evacuation efforts by terminating their services in the middle of a large scale relocation effort. These surveys clearly demon-strate the need to consider the potential effects of multiple group membership, espe-cially family ties, on the behavior of designated emergency personnel, particularly during nuclear emergencies.

TABLE 10 BEHAVIORAL INTENTIONS OF SELECTED EMERGENCY PERSONNEL[a]

Behavioral Response	Firemen[b] (n = 291)	Bus Drivers[b] (n = 246)	Teachers[c] (n = 226)
Assist With Evacuation	21	24	66.7
Make Sure Family Is Safe	68	73	23.5
Leave the Area Immediately	1	3	1.4
Do Something Else or Uncertain	11	—	8.5

[a]Data in percentages.
[b]D. J. Zeigler and Social Data Analysts 1982b.
[c]Johnson 1983

Perceptions of Nuclear Hazards

Why should the public response to nuclear disasters be different from the response to other types of disasters? The question may be answered most simply by pointing out that behaviors differ because perceptions differ. The catastrophic potential of nuclear power seems to color the perception of what has been an energy source responsible for not a single off-site death in the United States. As the *Washington Post* (January 28, 1982:A24) reminded its readers in a defense of the nuclear option: One American coal miner dies every other working day and in an average year 15,000 suffer disabling injuries. Why, therefore, are the risks of coal-fired generating plants accepted without major protest while nuclear generating plants arouse widespread controversy? Slovic and colleagues (1979) found that whereas nuclear power is perceived as involuntary, unknown to those exposed or to science, uncontrollable, unfamiliar, potentially catastrophic, severe, and dreaded, non-nuclear methods of generating electricity are perceived as more familiar, more voluntary, less catastrophic, and less dreaded. In their sample survey of the League of Women Voters, nuclear power was ranked as more dreaded than any of the other 29 activities or technologies on the list. Obviously, some segments of the public deem other factors more important than fatality or injury rates. Slovic and others (1980:190-192) have also found that the experts' assessments of risk were very closely correlated with the actuarial risk of death, whereas risk assessments among lay people were only moderately related to death rates. Their research seems to indicate that while technical experts evaluate risk by "back-casting," projecting the future from the past, the public is more prone to evaluate risk by "fore-casting," acknowledging that with new technologies the past is not always prologue. After asking their sample to estimate the number of deaths from various technologies during an average year, Slovic and colleagues (1980:192-193) asked the following question: "How many times more deaths would occur if next year were particularly disasterous rather than average?" Almost 40 percent of the respondents expected more than 10,000 fatalities and more than 25 percent expected more than 100,000 fatalities if next year were a disasterous year in the nuclear industry. Perhaps no other statistics reveal as dramatically the perceived "diaster potential" of nuclear power when compared with other technologies. These authors (Slovic *et al.* 1980: 208) also suggested that the gap between expert and public evaluations of nuclear risk may apply to other low probability/high consequence hazards such as LNG and recombinant DNA research. In short, "beliefs about the catastrophic nature of nuclear power are a major determinant of public opposition" (Slovic *et al.* 1980:207-208).

Another explanation for the fear of nuclear power may be bound up in its history. Cook (1982:4) documented how the federal government since World War II "failed at several times and places in its responsibility to protect the public from the adverse effects of the nuclear activities it was promoting." Lifton (1976) related the fear of nuclear power to the "terrible death" imagery associated with the use of nuclear weapons. Looming large in the collective public memory are the deaths and human suffering caused by bombing of Hiroshima and Nagasaki (Finch 1981; Thurlow 1982). Recent studies have also related abnormally high cancer incidence in Nevada to below ground weapons tests conducted there after the war (Lyon *et al.* 1979; Wasserman 1982). There has not been a history of responsible management of nuclear technology.

An additional explanation for the exaggerated behavioral reactions to nuclear emergencies may be the unique character of the disaster agent, ionizing radiation. While radioactivity is known to be potentially lethal, carcinogenic, and mutagenic, it is not visible or otherwise sensible. To quote Erikson (1982:56):

> *Most emergencies, whether they result from acts of God (such as floods, storms, or earthquakes) or acts of men (such as accidental explosions or deliberate bombings), have a clear beginning and a clear ending. Sooner or later the flood waters recede, the winds abate, the smoke clears, the bombers leave; an "all clear" is sounded both literally and figuratively to indicate that the incident is over and the source of danger is gone. But when an invisible threat hangs in the air or is lodged in the tissues of the body for an indeterminant amount of time, the survivors have no sure way of knowing how much damage has been done or is yet to be done, the event is never quite over. The cause for alarm never quite disappears.*

The invisibility of what is truly a mysterious disaster agent may help to explain the over-reaction of TMI area residents in 1979 and the high rates of evacuation projected on Long Island. In most disasters, direct sensory evidence is available to the public at large, emergency personnel, and the information media. In a nuclear disaster, only indirect evidence amassed from remote control panels and dosimeters is available, and most of that is interpretable only by experts. Whereas the residents of a flood plain can corroborate the reports of government and the news media through personal observation of rising waters and the information network of family and friends, the residents of an area threatened by a nuclear mishap have absolutely no access to primary evidence. They must instead depend on utility company and government reports as secondary sources of information (Brunn *et al.* 1979:35-36). These contrasts in the flow of information are diagrammed in Figure 26. In the event of a natural disaster, government officials, the news media, and the public all have access to data acquired through the human senses. At each stage in the flow of information, reported events can be directly compared with events as they are occurring, at which time incon- sistencies can be resolved. In the event of a nuclear disaster, the news media and the public have access to only secondary sources of information and even government

A. THE ACQUISITION OF INFORMATION IN THE
EVENT OF A NATURAL DISASTER

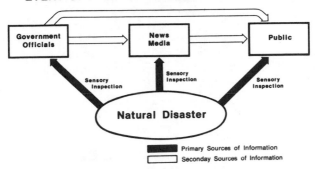

B. THE ACQUISITION OF INFORMATION IN THE
EVENT OF A NUCLEAR DISASTER

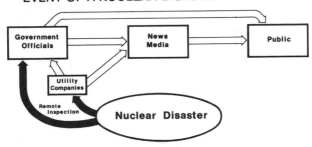

FIGURE 26   The Acquisition of Information in the Event of Natural and
Nuclear Disasters (Brunn, Johnson and Zeigler 1979:26)

officials must depend in large part on secondary data made available by the utility
companies. Sensory data are not available for the identification and evaluation of a
nuclear accident so that almost all information is acquired through remote inspection
with instruments that monitor the internal behavior of the reactor vessel or with
dosimeters that measure the magnitude of invisible radiation. Given the threat posed
by a severe nuclear accident, the existence of conflicting reports about conditions
during a nuclear emergency, and the lack of direct sensory evidence, it seems entirely
reasonable that over-reaction would be more prevalent than during natural disasters or
civil disorders. With 145 nuclear reactors operable, under construction, or on order in
the United States and 518 worldwide (American Nuclear Society 1983), there is an
undisputed need to understand both the reality of nuclear risk and the public response
to perceptions of the nuclear threat.

6

Technological Hazards of the Future

Technology has led to substantial improvements in social well-being and the quality of life for many segments of our global society. It has also vested upon people, places, and environments many of the negative externalities of its application. As the scale of technology has grown, so has the scale, catastrophic potential, and fear of technological hazards. "The place of natural threats is being taken by hazards which are increasingly rooted in the use and misuse of technology," noted Foster (1980:2). Dynes (1982:128) calls technological hazards "the natural disasters of the future." The questions which we ask ourselves about assaults originating in the technological environment have grown immeasureably. As we have come to question whether the benefits of a particular technology are worth the risks, the field of risk analysis has come into its own. Although geographers provided one of the two main lines of investigation from which risk analysis evolved (Otway and Thomas 1982:70), the spatial perspective is often lacking in the definition of risks and benefits. Inclusion of the geographic dimension often clarifies the issues and suggests the scale at which decisions concerning the acceptance, control, or rejection of technology should be made. In our increasingly pluralistic society, social movements, interest groups, pressure groups, and lobbies often perceive issues of risk, equity, and social justice differently (Covello *et al.* 1982:55). Often neglected is the fact that location in space strongly influences perceptions of these same issues. Just as the citizens of Alpena, Michigan, perceive the hazards of nuclear waste disposal differently because of their geography, the people of Western Europe perceive the hazards of the nuclear arms race differently from North Americans because of their global location. Bunge (1982) has called Europe the "walnut in the nuclear nut cracker" since it is located between the two major super powers and therefore a likely field on which a battle between the U.S. and the Soviet bloc would be fought.

Who is to decide whether radioactive wastes will be disposed of in Alpena? Who is to decide whether a new generation of missiles will be deployed in Western Europe? Who is to decide whether the Shoreham nuclear power plant on Long Island will open? We profess to believe in democratic control of societal institutions, but we have yet to resolve at what scale the democratic control of technology should be exercised. What is good for the nation may not be good for everyone, at every place, within the nation.

These differences in perception may be patterned in geographic space and may result from proximity to or distance from actual and potential threats. When does the self-interest of the locality take precedence over the self-interest of a larger geographic entity? Like the issue of self-determination for all peoples, we believe in the democratic control of technology, but at what scale?

Underlying our increasing concern about technology's negative physical, social, and economic side-effects is our scepticism that future technological advances can be expected to result in the same quantum leaps in quality of life that past technologies have made possible. Without a clear appreciation of technological benefits and with an enhanced awareness of technological risks, debates over technological advances often turn into burning social issues. Not all innovations are viewed with suspicion, however. Technologies which seem to trigger widespread emotional responses are those with catastrophic potential, those which are controlled by technocratic elites, and those which have been the subjects of debates among experts. If potential consequences of a technology are limited and risks are perceived to be manageable, a divided scientific community will make little difference. But as the consequences approach catastrophic proportions and as the risks are perceived to be uncontrollable, the more important does a schism in the scientific community become. Even a minority of scientific mavericks questioning a low probability/high consequence event can precipitate a major public debate.

We began this monograph by noting three hazards which demand our attention during the waning years of the twentieth century. All three owe their existence to the qualitative changes in technology that have taken place since World War II. The threats of toxic chemicals, nuclear power generation, and all-out nuclear war may be the last great hazards of the industrial era and its production-oriented and combative mindset. Many have pointed out that large-scale technology seems to have reached its social limits and is now being outmoded by the small-scale technologies of the information-intensive, post-industrial era. Whether this is true or not, the nuclear plants which are already scheduled to come on line will remain in operation well into the next century. In retrospect they may be viewed as transition energy sources to a future of fusion energy or renewable resources. For now though, they present society with a realm of hazards that must be better understood from both a technical and human perspective. Similarly the abundance of chemicals which we have thoughtlessly disposed of will continue to surface for scores of years, and future generations will pick up the tab for past errors in judgment. At three minutes to midnight on the Atomic Scientists' clock, the specter of nuclear war also seems ever-threatening. The thermo-nuclear sword of Damocles seems unlikely to be beaten into plowshares just because we have passed into an era of information-intensive technology. As Godet (1983:261) pointed out, "Technical progress and cultural change are accelerating, but in parallel there are no changes taking place in society, in institutions, or in individual persons to face up to such changes."

These hazards and most others discussed herein have resulted from releases of energy or materials. As society moves into a high-technology future, communications and information-processing industries will occupy a larger percentage of the labor force and extensively change our lives. (Fiegenbaum and McCorduck 1983). The hazards of the future may therefore result from the releases of information, or its opposite, the control and monopoly of information by irresponsible "hackers," white-collar criminals, or power-hungry reprobates. We see our privacy threatened, our confidential records less secure, and fear the unauthorized uses of computers (Turn and Ware 1975). Some people see their job being outmoded; others wonder whether the information economy will be able to gainfully employ every person who wants to work; some worry about the impact of artificial intelligence and Knowledge Information Processing Systems (KIPS) on their own identity as human beings; and still others worry about becoming overly dependent on computers and thus vulnerable to computer breakdowns and mistakes. Computer crime laws have been passed in seventeen states and what is evolving is a broader concept of "information crime statutes" (Parker 1983:240-244). Parker (1983:267) contends that the security of electronic money, the volume of which is growing, "is going to become an important part of national security as the economies of developed countries become increasingly reliant on EFT [electronic funds transfer] and the interconnection of systems." The nations of the developing world, on the other hand, may be concerned about the hazards of not being able to tie into the global information economy because of the barriers to diffusion which have been erected to block the North-to-South flow of technology and the information which powers it.

Information and the ease with which it overcomes the friction of distance have been cited by some as a decentralizing and democratizing social force. Lest one place too much hope in the democratizing influence of information technologies, however, it may be instructive to note how modern communications systems have helped to centralize political and economic power by permitting the growth of ever larger social institutions (Mandeville 1983:69). Multi-national corporations can operate at the global scale only by using communications technologies to assemble and coordinate large volumes of information. And, who could doubt that the role of the federal government in Washington would have grown to usurp so many state and local powers were it not for the bureaucracy's ability to communicate so easily with all parts of the United States and to assemble and manipulate multiple nationwide data bases.

Physicists have unleashed the power of the atom and now we debate the hazards of nuclear energy and nuclear war. Chemists have synthesized toxic chemicals unknown in nature and now we haggle over what to do about toxic waste dumps, pesticide residues in food, and lethal compounds which are virtually indestructable. Biologists have now embarked on an era of recombinant DNA research, gene splicing, and extended (even unlimited?) longevity, which may be responsible for the Three Mile Islands and Love Canals of the future. At the same time, however, genetic research is most likely to reveal the causes of cancer, multiple sclerosis, and Huntington's disease; genetic screening may become a widely used method of identifying the diseases to which an individual is prone; and genetic manipulation may be the cure for Tay-Sachs disease and sickle cell anemia. Are botanical gene-splicing experiments aimed at creating "super plants" guaranteed to benignly increase agricultural productivity, or are they tantamount to "ecological roulette" as Rifkin (1983) contends? Technology has given us increasing control over nature, but what gives us control over technology? When does technological choice become technological addiction? Where are we on the continuum between technological determinism and technological possibilism?

The scale of technology has grown and with it human dependence on complex, computer-controlled lifelines that some feel have deprived the individual of self-reliance and control over the future. Just as there are hazards in rejecting technological advances, there are hazards in overdependence on technology. Think for a moment what the ultimate in technological dependence would be: Synthetic food and recycled water; intelligent electrodes that put us to sleep at night and wake us up in the morning after sensing that we have received the optimal amount of rest; drugs to improve memory, reasoning abilities, even creativity; children whose sex and genetic endowments have been chosen from a gene bank; simulators combined with mind-altering drugs that take us on vacation without leaving home; pleasure experienced through the artificial stimulation of the brain; financial transactions and record-keeping taken care of by a computer; decisions made by machines endowed with artificial intelligence which can learn from mistakes just as we can. One alternative future sees mankind as a technological marionette whose strings are pulled by anthropomorphic machines. Perhaps questions about the desirability of such a future have been responsible for the movement toward appropriate, small-scale technology that serves peoples' needs and wants, a technology that offers choices rather than one that determines options. At either end of the continuum there are certain to be technological hazards, but under a decentralized technology, those hazards would be more amenable to individual and community control. Appropriate technology helps us to do things for ourselves; "high" technology allows other people, the technocrats, to do things for us according to their rules and regulations. As Robertson (1978:594) has suggested, "One of the most exciting new frontiers of technology will be the development and diffusion of sophisticated tools designed to enable ever-increasing numbers of people to become self-reliant and self-realized," Naisbitt (1982:133-136) cites the change from institutional to self-help as one of the "megatrends" shaping our transformation into a new society. Decentralized technologies, such as those that support at-home medical diagnosis and care, make possible our quest for greater self-reliance.

What should the present and future role of geographers be in understanding hazards and deciding how they should be managed? First, geographers need to be active in defining hazard zones, in assigning risks to particular places within these areas, and in delineating the impact zones which would result from different accident scenarios. Second, based on the definition of these hazard zones, geographers should be able to suggest methods of altering the relationship between human settlement or transportation patterns and the threatening aspects of the technological environment. Whittaker et al. (1983), for instance, illustrates some of the problems involved in using risk analysis to manage land use patterns near toxic-gas pipelines and explores the difficulty in choosing between "worst-case" zoning and "average case" zoning. He summarizes the geographic dimension of the dilemma by noting: "the problem became one of assigning numerical values to the risk as a function of distance" (Whittaker et al. 1982:165). In short, geographers need to be involved in both the academic and applied aspects of anticipatory hazard management, in trying to minimize the potential for harm should an accident occur or chronic condition persist in the technological environment.

Third, geographers have a major contribution to make in the realm of emergency response planning. Laura Thomas (1982), who is involved with radiological emergency planning, has pointed out that the geographer is one of the few professionals who can approach the planning process with the ability to aggregate all of the data which is crucial to the evolution of a satisfactory plan. Demography, land use interpretation, map reading, transportation, meteorology, human behavior analysis, and cartography are necessary ingredients of both emergency planning and geographic training. Epstein and colleagues (1982:315) also made a good case for the contributions of the geographer with respect to the problem of aquifers contaminated by toxic wastes: "For contamination control programs to be successful, all aquifers must be mapped and analyzed; all sources of contamination must be located; all recharge areas must be found; and groundwater supplies must be monitored on a periodic basis."

Fourth, geographers need to be involved in both technology assessments and social and environmental impact statements. Every law designed to regulate or manage technology will have a spatial and ecological impact that is amenable to geographic analysis. In this regard, geographers may be able to suggest appropriate scales for looking at technological impacts and to suggest appropriate locations for technological installations in order to minimize human, economic, and environmental losses. Fifth, in assessing technological impacts, geographers need to capitalize on the tradition established by natural hazards researchers in emphasizing the role of perception of technological hazards. Sixth, geographers need to explore the role of technology in structuring alternative spatial futures. At any point in history, there is not one future, but many. One of the strongest influences on the future is society's choice of technology. To make wise choices, however, the public needs to be shown what options are available and what the possible consequences are. In the realm of communicating these options to the public, geographers and other social scientists could play a much more aggressive role than they have played in the past.

Bibliography

Ahern, W. R. 1980. "California Meets the LNG Terminal," *Coastal Zone Management Journal* 7:185-221.

Ajzen, I. and M. Fishbein. 1980. *Understanding Attitudes and Predicting Social Behavior.* Englewood Cliffs, NJ: Prentice-Hall.

Akey, D. S. , Editor. 1982. *Encyclopedia of Associations '83.* Detroit: Gale Research.

American Nuclear Society. 1983. "World List of Nuclear Power Plants," *Nuclear News* 26 (August):83-102.

Anderson, R. F. and M. R. Greenberg. 1982. "Hazardous Waste Facility Siting," *Journal of the American Planning Association* 48:204-218.

Bach, W. 1979. "Impact of Increasing Atmospheric CO_2 Concentrations on Global Climate," *Environment International* 2:215-228.

Baker, E. J. 1979. "Predicting Responses to Hurricane Warnings: A Reanalysis of Data From Four Studies," *Mass Emergencies* 4:9-24.

Baum, A. et al. 1983. "Natural Disaster and Technological Catastrophe," *Environment and Behavior* 15:333-354.

Beebe, G. W. 1982. "Ionizing Radiation and Health." *American Scientist* 70:35-44.

Behr, P. 1978. "Controlling Chemical Hazards," *Environment* 20, 6:25-29.

Berman, D. M. 1970. *Death on the Job: Occupational Health and Safety Struggles in the United States.* New York and London: Monthly Review Press.

Bick, T. and C. Hohenemser. 1979. "Target: Highway Risks I. Taking Individual Aim," *Environment* 21, 1:16-20, 37-40.

Bick, T. and R. E. Kasperson. 1978. "Pitfalls of Hazard Management," *Environment* 20, 8:30-42.

Bick, T. and R. W. Kates. 1979. "Target: Highway Risks II. The Government Regulators," *Environment* 21, 2:29-38.

Bluestone, B. and B. Harrison. 1982. *The Deindustrialization of America.* New York: Basic Books.

Blum, A. et al. 1978. "Children Absorb Tris-BP Flame Retardant from Sleepwear," *Science* 201:1020-1023.

Boehmer-Christiansen, S.A. 1983. "Dumping Nuclear Waste into the Sea," *Marine Policy* 7:25-36.

Brooks, R. 1971. "Human Response to Recurrent Drought in Northeastern Brazil," *Professional Geographer* 23:40-44.

Brown, M. 1981. *Laying Waste: The Poisoning of America by Toxic Chemicals.* New York: Washington Square Press.

Brunn, S. D. et al. 1980. "Locational Conflict and Attitudes Regarding the Burial of Nuclear Wastes," *East Lakes Geographer* 15:24-40.

Brunn, S. D., J. H. Johnson, Jr., and D. J. Zeigler. 1979. *Final Report on a Social Survey of Three Mile Island Area Residents.* East Lansing, Michigan: Department of Geography, Michigan State University, August.

Bullard, R. D. 1983. "Solid Waste Sites and the Black Houston Community," *Sociological Inquiry* 53:273-288.

Bunge, W. W. 1973. "The Geography of Human Survival," *Annals, Association of American Geographers* 63:275-295.

Bunge, W. W. 1982. *The Nuclear War Atlas.* Victoriaville, Quebec: Society for Human Exploration.

Burmaster, D. E. and R. H. Harris. 1982. "Groundwater Contamination: An Emerging Threat," *Technology Review* 85 (July):51-62.

Burness, H. S. 1981. "Risk: Accounting for an Uncertain Future," *Natural Resources Journal* 21:723-734.

Burton, I. and R. W. Kates. 1964. "The Perception of Natural Hazards in Resource Management," *Natural Resources Journal* 3.412-441.

Burton, I. et al. 1978. *The Environment as Hazard.* New York: Oxford University Press.

Carpenter, S. R. 1978. "Developments in the Philosophy of Technology in America," *Technology and Culture* 19:93-99.

Carson, R. L. 1962. *Silent Spring.* Boston: Houghton Mifflin.

Carter, M. W. and A. A. Moghissi. 1977. "Three Decades of Nuclear Testing," *Health Physics* 33 (July):55-71.

Chenault, W. W. et al. 1979. *Evacuation Planning in the TMI Accident.* Washington, DC: Federal Emergency Management Agency.

Christian, J. J. 1983. "Love Canal's Unhealthy Voles," *Natural History* 92 (October):8-16.

Church, A. M. and R. D. Norton. 1981. "Issues in Emergency Preparedness for Radiological Transportation Accidents," *Natural Resources Journal* 21 (October):757-771.

Clark, B. D. et al. 1980. *Environmental Impact Assessment: A Bibliography With Abstracts.* New York: R. R. Bowker.

Clark, W. C. 1977. "Managing the Unknown." pp. 109-142 in R. W. Kates (editor), *Managing Technological Hazard.* Boulder: University of Colorado Press.

Clark University Center for Technology, Environment and Development and Decision Research. 1979. *Project Summary: Improving the Societal Management of Technological Hazards.* Eugene, OR: Decision Research.

Cohen, B. 1981. "High Level Radioactive Waste," *Natural Resources Journal* 21:703-722.

Cole, G. A. and S. B. Withey. 1981. "Perspectives on Risk Perceptions," *Risk Analysis* 1:143-163.

Collingridge, D. 1980. *The Social Control of Technology.* London: Frances Pinter.

Commoner, B. 1971. *The Closing Circle: Nature, Man, and Technology.* New York: Alfred A. Knopf.

Congressional Research Service, Library of Congress. 1975. *Facts on Nuclear War.* Washington, DC: Government Printing Office.

Congressional Research Service, Library of Congress. 1977. *National Energy Transportation. Vol. 1: Current Systems and Movements.* Washington, DC: Government Printing Office.

Cook, E. 1982. "The Role of History in the Acceptance of Nuclear Power," *Social Science Quarterly* 63:3-15.

Cotgrove, S. 1975. "Technology, Rationality, and Domination," *Social Studies of Science* 5:55-78.

Cotgrove, S. 1982. *Catastrophe or Cornucopia?* New York: John Wiley.

Council on Environmental Quality. 1981. *Environmental Trends.* Washington, DC: Government Printing Office.

Council on Environmental Quality. 1982. *Environmental Quality 1981.* Washington, DC: Government Printing Office.

Council on Environmental Quality. 1983. *Environmental Quality 1982.* Washington, DC: Government Printing Office.

Covello, V. T. *et al.* **1982.** "Risk Analysis, Philosophy, and the Social and Behavioral Sciences," *Risk Analysis* 2 (June):53-58.

Cumming, R. B. 1981. "Is Risk Assessment a Science?" *Risk Analysis* 1 (March):1-4.

Curtis, C. 1983. "Radwaste Dumping Delayed," *Oceans* 16 (May):22-23.

Cutter, S. and K. Barnes. 1982. "Evacuation Behavior and Three Mile Island," *Disasters* 6:116-124.

Dardis, R. 1980. "Value of a Life: New Evidence From the Marketplace," *American Economic Review* 70:1077-1082.

Deveault, D. 1981. "World Views on Energy, No Entropy." *Bioscience* 31:605.

Dodge, R. 1982. "The Effects of Indoor Pollution on Arizona Children," *Archives of Environmental Health* 37 (May/June):151-155.

Drabek, T. E. 1969. "Social Processes in Disaster: Family Evacuation," *Social Problems* 16:336-349.

Drabek, T. E. and K. S. Boggs. 1968. "Families in Disaster: Reactions and Relatives," *Journal of Marriage and the Family* 30:443-451.

Drabek, T. E. and J. S. Stephenson II. 1971. "When Disaster Strikes," *Journal of Applied Social Psychology* 1:187-203.

Dynes, R. R. 1982. "The Accident at Three Mile Island," pp. 119-130 in D. L. Sills *et al.* (editors), *Accident at Three Mile Island: The Human Dimensions.* Boulder, CO: Westview Press.

Eckholm, E. P. 1977. *The Picture of Health: Environmental Sources of Disease.* New York: Norton.

Eckholm, E. P. 1978. *Disappearing Species: The Social Challenge.* Worldwatch Paper 22. Washington, DC: Worldwatch Institute.

Egan, J. R. and S. Arungu-Olende. 1980. "Nuclear Power for the Third World?" *Technology Review* 82 (May):46-55.

Ehrlich, P. and A. Ehrlich. 1981. *Extinction: The Causes and Consequences of the Disappearance of Species.* New York: Random House.

Ellul, J. 1964. *The Technological Society.* New York: Random House.

Ellul, J. 1980. *The Technological System.* New York: Continuum.

Epstein, S. S. *et al.* **1982.** *Hazardous Waste in America.* San Francisco: Sierra Club Books.

Erikson, K. T. 1976. *Everything in its Path: Destruction of Community in the Buffalo Creek Flood.* New York: Simon and Schuster.

Erickson, K. T. 1982. "Human Response in a Radiological Accident," In *The Indian Point Book.* Cambridge, MA and New York: Union of Concerned Scientists and New York Public Interest Research Group.

Etzioni A. and C. Nunn. 1974. "The Public Appreciation of Science in Contemporary America," *Daedalus* 103:191-205.

Farhar, B. C. *et al.* **1980.** "Public Opinion About Energy," *Annual Review of Energy* 5:141-172.

Federal Emergency Management Agency. 1980. *Report to the President: State Radiological Emergency Planning and Preparedness in Support of Commercial Nuclear Power Plants.* Washington, DC: Government Printing Office.

Ferkiss, V. C. 1974. *The Future of Technological Civilization.* New York: Braziller.

Fiegenbaum, E. A. and P. McCorduck. 1983. *The Fifth Generation: Artificial Intelligence and Japan's Computer Challenge to the World.* Reading, MA: Addison Wesley.

Finch, S. C. 1981. "Occurrence of Cancer in Atomic Bomb Survivors," In R. Adams and S. Cullen (editors), *The Final Epidemic: Physicians and Scientists on Nuclear War.* Chicago: Educational Foundation for Nuclear Science.

Fischhoff, B. 1979. "Behavioural Aspects of Cost-Benefit Analysis," pp. 270-283 in G. Goodman and W. D. Rowe (editors), *Energy Risk Management.* London: Academic Press.

Fischhoff, B. *et al.* **1978.** "Handling Hazards," *Environment* 20 (September):15-20, 32-37.

Fiske, S. T. *et al.* **1983.** "Special Issue: Images of Nuclear War," *Journal of Social Issues* 39:1-97.

Flynn, C. B. 1979. *Three Mile Island Telephone Survey: Preliminary Report on Procedures and Findings.* NUREG/CR-1093. Washington, DC: U.S. Nuclear Regulatory Commission.

Foster, H. D. 1976. "Assessing Disaster Magnitude: A Social Science Approach," *Professional Geographer* 28:241-247.

Foster, H. D. 1980. *Disaster Planning: The Preservation of Life and Property.* New York: Springer-Verlag.

Galligan, J. E. 1983. Personal Communication from the Upjohn Company. November 3, 1983.

Gibbons, J. H. 1983. "Technology Assessment Comes of Age," *Environment* 25 (January/February):28-31.

Gladwin, T. N. 1980. "Patterns of Environmental Conflict Over Industrial Facilities in the United States, 1970-78," *Natural Resources Journal* 20:243-274.

Glasstone, S. and P. J. Dolan (editors). 1977. *The Effects of Nuclear Weapons.* Washington, DC: Department of Defense and Department of Energy.

Glick, B. J. 1982. "The Spatial Organization of Cancer Mortality," *Annals, Association of American Geographers* 72 (December):471-481.

Godet, M. 1983. "Crisis and Opportunity — From Technological to Social Change," *Futures* 15:251-263.

Goldberg, E. D. 1981. "The Oceans as Waste Space: The Argument." *Oceanus* 24 (Spring):2-9.

Goldhaber, M. K. and J. E. Lehman. 1983. "Crisis Evacuation During The Three Mile Island Nuclear Accident: The TMI Population Registry," Paper presented at the 1982 annual meeting of the American Public Health Association, Montreal, Quebec (revised).

Goldman, R. *et al.* **1973.** "Dimensions of Attitudes Toward Technology," *Journal of Applied Psychology* 57:184-187.

Gordon, G. E. 1981. "Thermodynamics and Society," *Science* 211:1340-1341.

Graham, L. R. 1978. "Concerns About Science and Attempts to Regulate Inquiry," *Daedalus* 107:1-21.

Grahman, J. D. and J. W. Vaupel. 1981. "Value of a Life: What Difference Does It Make?" *Risk Analysis* 1 (March):89-95.

Greenberg, M. R. *et al.* **1978.** *Environmental Impact Statements.* Washington, DC: Association of American Geographers, Resource paper 78/3.

Greenberg, M. R. *et al.* **1980.** "Clues for Case Control Studies of Cancer in the Northeast Urban Corridor," *Social Science and Medicine* 14D:37-43.

Greene, M. R. *et al.* **1981.** "The March 1980 Eruptions of Mt. St. Helens: Citizen Perceptions of Volcano Threat," *Disasters* 5:49-66.

Hammond, A. L. 1975. "Ozone Destruction: Problem's Scope Grows, Its Urgency Recedes," *Science* 187:1181-1183.

Hammond, A. L. and T. H. Maugh II 1974. "Stratospheric Pollution: Threat to Earth's Ozone," *Science* 186:335-338.

Hans, J. M., Jr. and T. C. Sell. 1974. *Evacuation Risks — An Evaluation.* Las Vegas: U.S. Environmental Protection Agency.

Hardy, E.P. *et al.* **1968.** "Strontium-90 on the Earth's Surface," *Nature* 219:584-587.

Harris, R. C. and C. Hohenemser. 1978. "Mercury: Measuring and Managing the Risk," *Environment* 20 (November):25-36.

Harris, R. C. *et al.* **1978.** "Our Hazardous Environment," *Environment* 20 (September):6-15, 38-40.

Harvey, M. E. *et al.* 1979. "Cognition of a Hazardous Environment: Reactions to Buffalo Airport Noise," *Economic Geography* 55:263-286.

Heathcote, R. L. 1969. "Drought in Australia: A Problem of Perception," *Geographical Review* 59:175-194.

Hendrey, G. R. 1981. "Acid Rain And Gray Snow," *Natural History* 90 (February):58-65.

Hewitt, K. 1983. "Place Annihilation: Area Bombing and the Fate of Urban Places," *Annals, Association of American Geographers* 73:257-284.

Hewitt, K. and I. Burton. 1971. *The Hazardousness of a Place: A Regional Ecology of Damaging Events.* Toronto: University of Toronto Press.

Hohenemser, C. *et al.* 1983. "The Nature of Technological Hazard," *Science* 220:378-384.

Holdgate, M. W. *et al.* 1982. *The World Environment 1972-1982.* Dublin: Tycooly International.

Holdren, J. P. 1982. "Energy Hazards: What to Measure, What to Compare," *Technology Review* 85 (April):32-38,74-75.

Holdren, J. P. *et al.* 1980. "Environmental Aspects of Renewable Energy Sources," *Annual Review of Energy* 5:241-291.

Houts, P. S. and M. K. Goldhaber. 1981. "Psychological and Social Effects of the Population Surrounding Three Mile Island After the Nuclear Accident on March 28, 1979," pp. 151-164 in S. Majumdar (editor), *Energy, Environment and the Economy.* Harrisburg: Pennsylvania Academy of Science.

Houts, P. S. *et al.* 1980. "Extent and Duration of Psychological Distress of Persons in the Vicinity of Three Mile Island," *Proceedings, Pennsylvania Academy of Sciences* 54:22-28.

Isaksen, I. S. and F. Stordal. 1981. "The Influence of Man on the Ozone Layer: Readjusting the Estimates," *Ambio* 10:9-17.

Islam, S. 1983. "Red Alert on Corporate Crime," *South,* October:50-51.

Jakimo A. and I. C. Bupp. 1978. "Nuclear Waste Disposal: Not in My Backyard," *Technology Review* 80 (March/April):64-72.

Janis, I. L. 1962. "Psychological Effects of Warnings," in G. W. Baker and D. W. Chapman (editors), *Man and Society in Disaster.* New York: Basic Books.

Johnson, A. 1978. "Unnecessary Chemicals," *Environment* 20 (March):6-11.

Johnson, C. J. 1981. "Cancer Incidence in an Area Contaminated with Radionuclides Near a Nuclear Installation," *Ambio* 10:176-182.

Johnson, J. H., Jr. 1983. "Reactions of Public School Teachers to a Possible Accident at the Diablo Canyon Nuclear Power Plant," Unpublished Manuscript.

Johnson, J. H., Jr. and D. J. Zeigler. 1982. *Further Analysis and Interpretation of the Shoreham Evacuation Survey.* In Vol. 3 of the Suffolk County Radiological Emergency Response Plan (Draft), November, 1982.

Johnson, J. H., Jr. and D. J. Zeigler, 1983. "Distinguishing Human Responses to Radiological Emergencies," *Economic Geography* 59:386-402.

Johnson, J. H., Jr. and D. J. Zeigler, 1984. "A Spatial Analysis of Evacuation Intentions at the Shoreham Nuclear Power Station," pp. 279-301 in M. J. Pasqualetti and K. D. Pijawka (editors), *Nuclear Power: Assessing and Managing Hazardous Technology.* Boulder, CO: Westview Press.

Juenger, F. G. 1956. *The Failure of Technology.* Chicago: Henry Regnery.

Kahn, H. *et al.* 1976. *The Next 200 Years.* New York: William Morrow.

Kamlet, K. S. 1981. "The Oceans as Waste Space: The Rebuttal," *Oceanus* 24 (Spring):10-17.

Kaplan, S. and B. J. Garrick. 1981. "On the Quantitative Definition of Risk," *Risk Analysis* 1:11-27.

Karl, T. R. 1980. "A Study of the Spatial Variability of Ozone and Other Pollutants at St. Louis, Missouri," *Atmospheric Environment* 14:681-694.

Kasl, S. et al. 1981. "The Impact of the Accident at TMI on the Behavior and Well-Being of Nuclear Workers," *American Journal of Public Health* 71:472-495.

Kasperson, R. 1977. "Societal Management of Technological Hazards." pp. 51-79 in R. W. Kates 1977b.

Kasperson, R. et al. 1982. "Institutional Responses to Different Perceptions of Risk." pp. 39-46 in D. L. Sills et al. (editors), *Accident at Three Mile Island.* Boulder, CO: Westview Press.

Kates R. W. 1962. *Hazard and Choice Perception in Flood Plain Management.* Chicago: Department of Geography, University of Chicago.

Kates, R. W. 1977a. "Assessing the Assessors: The Art and Ideology of Risk Assessment," *Ambio* 6:247-252.

Kates, R. W. (editor). 1977b. *Managing Technological Hazards: Research Needs and Opportunities.* Boulder, CO: Institute of Behavioral Science, University of Colorado.

Kates, R. W. 1978. *Risk Assessment of Environmental Hazard.* Scientific Committee on Problems of the Environment.

Kearney, C. H. 1982. *Nuclear War Survival Skills.* Coos Bay, OR: NWS Research Bureau.

Keeney, R. B. et al. 1978. "Assessing the Risk of an LNG Terminal," *Technology Review* 81 (October):64-72.

Keillor, J. P. 1980. "The Hazards of Tank Ships and Barges Transporting Petroleum Products on the Great Lakes," *Coastal Zone Management Journal* 8:319-336.

Keisling, B. 1980. *Three Mile Island: Turning Point.* Seattle, WA: Veritas Books.

Killian, L. M. 1952. "The Significance of Multiple Group Membership in Disaster," *American Journal of Sociology* 57:309-314.

King, J. 1977. "A Science for the People," *New Scientists* 74:634-636.

King, J. 1980. "New Genetic Technologies: Prospects and Hazards," *Technology Review* 82 (February):57-61,64-65.

Kopecki, K. 1981. "The Case for Nuclear Energy," *International Social Science Journal* 23:481-494.

Krimsky, S. 1982. *Genetic Alchemy: The Social History of the Recombinant DNA Controversy.* Cambridge, MA: MIT Press.

Kunreuther, H. and J. W. Lathrop. 1981. "Siting Hazardous Facilities: Lessons from LNG," *Risk Analysis* 1:289-302.

LaPorte, T. and D. Metlay. 1975a. "Technology Observed: Attitudes of a Wary Public," *Science* 188:121-127.

LaPorte, T. and D. Metlay, 1975b. "Public Attitudes Toward Present and Future Technologies: Satisfactions and Apprehension," *Social Studies of Science* 5:373-398.

LaPorte, T. and D. Metlay. 1976. "Public Attitudes Toward Present and Future Technologies," *Ekistics* 42:165-168.

Lawless, E. W. 1977. *Technology and Social Shock.* New Brunswick, NJ: Rutgers University Press.

Laycock, G. 1976. "A Dam is Not Difficult to Build Unless it is in the Wrong Place," *Audubon* 78 (November):132-135.

Lee, N. 1983. "Environmental Impact Assessment: A Review," *Applied Geography* 3:5-27.

Levine, A. G. 1982. *Love Canal: Science, Politics, and People.* Lexington, MA: Lexington Book.

Lifton, R. J. 1964. "Psychological Effects of the Atomic Bomb in Hiroshima: The Theme of Death," pp. 152-193 in G. H. Grosser et al. (editors), *The Threat of Impending Disaster.* Cambridge, MA: MIT Press.

Lifton, R. J. 1976. "Nuclear Energy and the Wisdom of the Body," *Bulletin of the Atomic Scientists* 32:16-20.

Likens, G. E. and F. H. Bumann. 1974. "Acid Rain: A Serious Regional Environmental Problem," *Science* 184:1176-1179.

Liverman, D. and J. P. Wilson. 1981. "The Mississauga Train Derailment and Evacuation, 10-16 November 1979," *Canadian Geographer* 25:365-375.

Lyon, J. L. et al. 1979. "Childhood Leukemias Associated With Fallout From Nuclear Testing," *New England Journal of Medicine* 300:397-402.

Macgill, S. M. and D. J. Snowball. 1983. "What Use Risk Assessment?" *Applied Geography* 3:171-192.

Macleod, G. K. 1981. "Some Public Health Lessons From Three Mile Island: A Case Study in Chaos," *Ambio* 10:18-23.

Macy, J. R. 1983. *Despair and Personal Power in the Nuclear Age.* Philadelphia: New Society Press.

Mandeville, T. 1983. *"The Spatial Effects of Information Technology,"* *Futures* 15 (February):65-72.

Marino, A. A. and R. A. Becker. 1978. "High Voltage Lines: Hazard at a Distance," *Environment* 20 (November):6-15.

Martin, D. 1980. *Three Mile Island: Prologue or Epilogue?* Cambridge, MA: Ballinger.

Maslow, A. H. 1970. *Motivation and Personality.* New York: Harper and Row.

Mason, P. F. 1971. "Man, the Urban Environment, and Massive Industrial Accidents: A Geographical Perspective," *Journal of Geography* 70:328-330.

Mason, P. F. 1972. "Some Spatial Implications of a Massive Industrial Accident: The Case of Nuclear Power Plants," *Professional Geographer* 24:233-236.

Mayer, J. D. 1981. "Geographical Clues About Multiple Sclerosis," *Annals, Association of American Geographers* 71:28-39.

Maxwell, C. 1982. "Hospital Organizational Response to the Nuclear Accident at Three Mile Island," *American Journal of Public Health* 72:275-279.

May, W. W. 1982. "$$ for Lives: Ethical Considerations in the Use of Cost/Benefit Analysis by For-Profit Firms," *Risk Analysis* 2 (March):35-46.

Mazur, A. 1973. "Opposition to Technological Innovation," *Mionerva* 11:243-262.

McGrath, P. E. 1974. *Radioactive Waste Management: Potentials and Hazards From a Risk Point of View.* Karlsruhe, Germany: U.S.-EURATOM Fast Reactor Exchange Program, Report EURFNR-1204.

Medvedev, Z. A. 1979. *Nuclear Disaster in the Urals.* New York: Norton.

Mileti, D. and E. M. Beck. 1975. "Communication in Crisis: Explaining Evacuation Symbolically," *Communication Research* 2:24-49.

Miller, M. W. and G. E. Kaufman. 1978. "High Voltage Overhead," *Environment* 20:6-16.

Mingst, K. A. 1982. "Evaluating Public and Private Approaches to International Solutions to Acid Rain Pollution," *Natural Resources Journal* 22:5-20.

Minowa, M. et al. 1981. "Geographical Distribution of Lung Cancer Mortality and Environmental Factors in Japan," *Social Science and Medicine* 15D:225-231.

Mitchell, B. 1979. "Natural Hazards," pp. 201-227 in *Geography and Resource Analysis.* London: Longman.

Mitchell, J. K. 1974a. *Community Response to Coastal Erosion: A Study of Conditions on the Atlantic Shore.* Chicago: Department of Geography, University of Chicago.

Mitchell, J. K. 1974b. "Natural Hazards Research," pp. 311-341 in I. R. Manners and M. W. Mikesell (editors), *Perspectives on the Environment.* Washington, DC: Association of American Geographers.

Moore, H. E. et al. 1963. *Before the Wind: A Study of the Response to Hurricane Carla.* Washington, DC: National Academy of Sciences/National Research Council, Disaster Study No. 19.

Moser, L. J. 1979. *The Technology Trap: Survival in a Man-Made Environment.* Chicago: Nelson-Hall.

Munafo, J. L. 1978. "Energy Transmission: A Social Geographic Disaster?" *Transition* 8 (Summer):31-33.

Naisbitt, J. 1982. *Megatrends.* New York: Warner Books.

National Academy of Sciences, National Research Council. 1975. *Long-Term Worldwide Effects of Multiple Nuclear-Weapons Detonations.* Washington, DC: National Academy of Sciences.

National Academy of Sciences, National Research Council. 1983. *Acid Deposition: Atmospheric Processes in Eastern North America.* Washington, DC: National Academy of Sciences.

National Safety Council. 1981. *Accident Facts.* Chicago: National Safety Council.

Nelkin, D. 1975. "The Political Impact of Technical Expertise," *Social Studies of Science* 5:35-54.

Nelkin, D. 1983. "Workers at Risk," *Science* 222:125.

Nelkin, D. and S. Fallows. 1978. "The Evolution of the Nuclear Power Debate: The Role of Public Participation," *Annual Review of Energy* 3:275-312.

Nelkin, D. and M. Pollak. 1979. "Public Participation in Technological Decisions: Reality or Grand Illusion?" *Technology Review* 81 (September):54-65.

Nisbet, I. C. 1974. "Salt on the Earth," *Technology Review* 76 (May):6-7.

Norman, C. 1981. *The God That Limps.* New York: W. W. Norton.

"Occupational Hazards." 1983. *Development Forum.* January-February 1983:6.

Office of Technology Assessment. 1979. *The Effects of Nuclear War.* Washington, DC: Office of Technology Assessment.

Okrent, D. 1980. "Comment on Societal Risk," *Science* 208:372-375.

Okrent, D. and D. W. Moeller. 1981. "Implications for Reactor Safety of the Accident at Three Mile Island, Unit 2," *Annual Review of Energy* 6:43-88.

Olson, S. H. 1979. "An Ecology of Workplace Hazards," *Economic Geography* 55:287-307.

Openshaw, S. and P. Steadman. 1982. "On the Geography of a Worst Case Nuclear Attack on the Population of Britain," *Political Geography Quarterly* 1:263-278.

Openshaw, S. and P. Steadman. 1983a. "The Bomb — Where Will the Survivors Be?" *Geographical Magazine* 55 (June):293-296.

Openshaw, S. and P. Steadman. 1983b. "The Geography of Two Hypothetical Nuclear Attacks on Britain," *Area* 15:193-201.

Opinion Research Corporation. 1974. *Attitudes of the U.S. Public Towards Science and Technology.* Caravan Surveys. Washington, DC: National Science Foundation.

O'Riordan. T. 1979. "The Scope of Environmental Risk Management," *Ambio* 8:260-264.

Otway, H. and K. Thomas. 1982. "Reflections on Risk Perception and Policy," *Risk Analysis* 2 (June): 69-82.

Parker, D. B. 1983. *Fighting Computer Crime.* New York: Charles Scribner's Sons.

Pasqualetti, M. J. 1983. "Nuclear Power Impacts: A Convergence/Divergence Schema," *Professional Geographer* 35:427-436.

Patterson, W. C. 1979. "Austria's Nuclear Referendum," *Bulletin of the Atomic Scientists* 35,1:6-7.

Pearce, D. 1983. "Ethics, Irreversibility, Future Generations and the Social Rate of Discount," *International Journal of Environmental Studies* 21:67-86.

Perry, R. W. 1979a. "Incentives for Evacuation in Natural Disasters," *Journal of the American Institute of Planners* 45:440-447.

Perry, R. W. 1979b. "Evacuation Decision-Making in Natural Disasters," *Mass Emergencies* 4 (March):25-38.

Perry, R. W. et al. 1981. *Evacuation Planning in Emergency Management.* Lexington, MA: Lexington Books.

Peterson, J. 1983. *The Aftermath: The Human and Ecological Consequences of Nuclear War.* New York: Pantheon.

Pijawka, D. and J. Chalmers. 1983. "Impacts of Nuclear Generating Plants on Local Areas," *Economic Geography* 59 (January):66-80.

Porter, A. L. et al. 1980. *A Guidebook for Technology Assessment and Impact Analysis.* New York: North Holland.

The President's Commission on the Accident at Three Mile Island. 1979. *The Need for Change: The Legacy of TMI.* Report of the President's Commission on the Accident at Three Mile Island. Washington, DC: U.S. Government Printing Office.

Preston, V. et al. 1983. "Adjustment to Natural and Technological Hazards: A Study of an Urban Residential Community," *Environment and Behavior* 15 (March):143-164.

Pyle, J. A. and R. G. Derwent. 1980. "Possible Ozone Reductions and UV Changes at the Earth's Surface," *Nature* 86 (24 July):373-375.

Quarantelli, E. L. (editor). 1978. *Disasters: Theory and Research.* Beverly Hills, CA: Sage Publications.

Quarantelli, E. L. et al. 1978. "Initial Findings from a Study of Socio-Behavioral Preparation and Planning for Acute Chemical Hazard Disasters," *Journal of Hazardous Materials* 3 (February):77-90.

Ramo, S. 1983. *What's Wrong With Our Technological Power in the Nuclear Age.* Philadelphia: New Society Press.

Rasmussen, N. C. 1981. The Application of Probabilistic Risk Assessment Techniques to Energy Technologies," *Annual Review of Energy* 6:123-138.

Rayner, J. F. 1953. *Hurricane Barbara: A Study of the Evacuation of Ocean City, Maryland.* Washington, DC: National Academy of Sciences/National Research Council.

Rees, J. D. 1970. "Paricutin Revisited: A Review of Man's Attempts to Adapt to Ecological Changes Resulting from Volcanic Catastrophe," *Geoform* 4:7-25.

Regenstein, L. 1982. *America the Poisoned.* Washington, DC: Acropolis Books.

Reissland, J. and V. Harries. 1979. "A Scale for Measuring Risk," *New Scientist* 83 (13 September):809-811.

Revelle, R. 1982. "Carbon Dioxide and World Climate," *Scientific American* 247,2:35-44.

Rhodes, S. L. and P. Middleton. 1983. "The Complex Challenge of Controlling Acid Rain," *Environment* 25(May):6-9,31-38.

Rich, L. D. 1970. *The Coast of Maine.* New York: Thomas Y. Crowell.

Richards, B. 1980. "Sierra Leone Rejects Chemical Waste Plan," *Washington Post* February 22, 1980:A10.

Rifkin, J. and T. Howard. 1981. *Entropy: A New World View.* New York: Bantam Books.

Rifkin, J. 1983. *Algeny.* New York: Viking Press.

Robertson, J. 1978. "Technology — Master or Servant?" *New Scientist* 80:594.

Rosenfield, A. et al. 1983. "The Food and Drug Administration and Medroxyprogesterone Acetate," *Journal of the American Medical Association* 249:2922-2928.

Saarinen, T. F. 1966. *Perception of the Drought Hazard on the Great Plains.* Chicago: Department of Geography, University of Chicago.

Sandia Laboratories, Inc. 1982. *Technical Guidance for Siting Criteria Development.* NUREG/CR-2239. Washington, DC: Nuclear Regulatory Commission.

Schulze, W. D. et al. 1981. "The Social Rate of Discount for Nuclear Waste Storage: Economics or Ethics? *Natural Resources Journal* 21:809-832.

Schumacher, E. F. 1973. *Small is Beautiful: Economics as if People Mattered.* New York: Harper and Row.

Shapiro, F. C. 1981. *Radwaste: A Reporter's Investigation of a Growing Problem.* New York: Random House.

Sheridan, T. B. 1980. "Human Error in Nuclear Power Plants," *Technology Review* 82 (February):22-33.

Shue, H. 1981. "Exporting Hazards," *Ethics* 91:579-606.

Sims, J. and D. D. Baumann. 1972. "The Tornado Threat: Coping Styles of the North and South," *Science* 176:1386-1392.

Sims, J. and T. F. Saarinen. 1969. "Coping With Environmental Threat: Great Plains Farmers and the Sudden Storm," *Annals, Association of American Geographers* 59:677-686.

Sims, J. H. and D. D. Baumann. 1983. "Educational Programs and Human Responses to Natural Hazards," *Environment and Behavior* 15:165-189.

Singer, M. 1977. "Scientists and the Control of Science," *New Scientist* 74:631-634.

Sjostrom, H. and R. Nilsson. 1972. *Thalidomide and the Power of the Drug Companies.* Hammondsworth, UK:Penguin.

Slovic, P. et al. 1979. "Rating the Risks," *Environment* 21 (April):14-20,36-39.

Slovic, P. et al. 1980. "Facts and Fears: Understanding Perceived Risk," pp. 181-214 in R.C. Schwing and W.A. Albers, Jr., (editors), *Societal Risk Assessment.* New York: Plenum Press.

Slovic, P. et al. 1982. "Psychological Aspects of Risk Perception," pp. 11-20 in D. L. Sills (editor), *Accident at Three Mile Island.* Boulder, CO: Westview Press.

Slovic, P. and B. Fischhoff. 1983. "How Safe is Safe Enough?" pp. 112-150 in C. A. Walker et al., (editors), *Too Hot to Handle?* New Haven, Conn.: Yale University Press.

Smalley, R. D. 1980. "Risk Assessment: An Introduction and Critique," *Coastal Zone Management Journal* 7:133-162.

Smith, G. 1976. "Beggaring Thy Neighbor or Thyself: Risk Factors in the Siting of Liquified Natural Gas Terminals," *Proceedings, Association of American Geographers* 8:139-143.

Smith, J. S. and J. H. Fisher. 1981. "Three Mile Island: The Silent Disaster," *Journal of the American Medical Association* 245:1656-1659.

Social Data Analysts, Inc. 1982a. *Attitudes Toward Evacuation: Reactions of Long Island Residents to a Possible Accident at the Shoreham Nuclear Power Plant.* A Report Prepared for Suffolk County, NY.

Social Data Analysts, Inc., 1982b. *Responses of Emergency Personnel to a Possible Accident at the Shoreham Nuclear Power Plant.* A Report Prepared for Suffolk County, NY.

Stallen, P. J. M. 1980. "Risk of Science or Science of Risk?" pp. 131-148 in J. Conrad (editor), *Society, Technology, and Risk Assessment.* London and New York: Academic Press.

Stanley, M. 1978. *The Technological Conscience.* New York: Free Press.

Starr, C. 1969. "Social Benefit versus Technological Risk," *Science* 165:1232-1238.

Starr, C. et al. 1976. "Philosophical Basis for Risk Analysis," *Annual Review of Energy* 1:629-662.

Starr, C. and C. Whipple. 1981. "Coping with Nuclear Power Risks: The Electric Utility Incentives," Paper presented at the International Workshop on the Analysis of Actual vs. Perceived Risks, Washington, DC.

Stengel, R. 1982. "Freezing Nukes, Banning Bottles," *Time,* November 15, 1982:34.

Sternglass, E. J. 1969. Infant Mortality: The Nuclear Culprit," *Medical Tribune,* July 21, 1969:15.

Sterrett, F. S. 1977. "Drinkable But . . ." *Environment* 19 (December):28-36.

***Sunday Patriot News.* 1982.** "Restart? No!" May 16, 1982:A4.

Sussman, B. 1983. "The Washington Post Poll," *The Washington Post National Weekly Edition,* December 5:12.

Taviss, I. 1972. "A Survey of Popular Attitudes Toward Technology," *Technology and Culture* 13:606-621.

Tausend, S. 1983. "Decision-Making and the Distribution of Depo-Provera." Unpublished paper, University of Kentucky, Department of Sociology.

Terman, C. R. and R. J. Huggett. 1980. "Occurrence of Kepone in White-Footed Mice on Jamestown Island, Virginia," *Environment International* 3:307-310.

Thomas L. 1982. "The Importance of Geography to Radiological Emergency Planning." Paper Presented at the Applied Geography Conference, College Park, Maryland.

Thurlow, S. 1982. "Nuclear War in Human Perspective: A Survivor's Report," *American Journal of Orthopsychiatry* 54:638-645.

Tichener, P. J. et al. 1971. "Environment and Public Opinion," *Journal of Environmental Education* 2 (Summer):38-42.

Toffler, A. 1970. *Future Shock.* New York: Random House.

Toffler, A. 1980. *The Third Wave.* New York: William Morrow.

Toth, R. C. 1981. "Scientists Take Another Look at SST," *The Virginian-Pilot,* February 15:K7.

Trabalka, J. R. et al. 1980. "Analysis of the Soviet Nuclear Accident," *Science* 209:345-353.

Turn, R. and W. H. Ware. 1975. "Privacy and Security in Computer Systems," *American Scientist* 63:196-203.

Turner, B. A. 1978. *Man-made Disasters.* London: Wykeham.

Tversky, A. and D. Kahneman. 1974. "Judgment Under Uncertainty: Heuristics and Biases," *Science* 185:1124-1131.

Union of Concerned Scientists. 1977. *The Risks of Nuclear Power Reactors: A Review of the NRC Reactor Safety Study.* Cambridge, MA: Union of Concerned Scientists.

United Nations Environment Programme. 1982. *The Health of the Oceans.* Regional Sea Reports and Studies No. 16. New York: United Nations.

U.S. Arms Control and Disarmament Agency. 1975. *Worldwide Effects of Nuclear War. Congress.* Washington, DC: Government Printing Office.

U.S. Arms Control and Disarmament Agency. 1976. *15th Annual Report to the Congress.* Washington, DC: Government Printing Office.

U.S. Congress. House of Representatives. 1978. "U.S. Export of Banned Products," *Hearings:* July 11, 12, and 13, 1978. Washington, DC: Government Printing Office.

U.S. Congress. House of Representatives and Senate. 1980. "Risk/Benefit Analysis in the Legislative Process," *Hearings:* July 24 and 25, 1979. Washington, DC: Government Printing Office.

U.S. Department of Transportation. 1982. *Report on Traffic Accidents and Injuries for 1979-1980* (COMSIS Corp.). Washington, DC: Government Printing Office.

U.S. Department of Transportation. 1983. *Fatal Accident Reporting System 1981.* Washington, DC: Government Printing Office.

U.S. Environmental Protection Agency. 1980. *Hazardous Materials Incidents Reported to U.S. Environmental Protection Agency Regional Offices From October 1977 Through September 1979.* Washington, DC: U. S. Environmental Protection Agency.

U.S. Environmental Protection Agency. 1983. *Can We Delay a Greenhouse Warming?* Washington, DC: Government Printing Office.

U.S. Nuclear Regulatory Commission. 1975. *Reactor Safety Study: An Assessment of Accident Risks in U.S. Nuclear Power Plants* (by N. C. Rasmussen). WASH-1400. Washington, DC: Nuclear Regulatory Commission.

U.S. Nuclear Regulatory Commission. 1977. *Socio-Economic Impacts: Nuclear Power Station Siting.* State College, PA: Policy Research Associates.

U.S. Nuclear Regulatory Commission. 1978a. *Planning Basis for the Development of State and Local Government Radiological Emergency Response Plans in Support of Light Water Nuclear Power Plants.* NUREG-0396. Washington, DC: Government Printing Office.

U.S. Nuclear Regulatory Commission. 1978b. *Risk Assessment Review Group Report to the U.S. Nuclear Regulatory Commission.* Washington, DC: Nuclear Regulatory Commission.

U.S. Nuclear Regulatory Commission. 1979. *Population Dose and Health Impact of the Accident at Three Mile Island.* Washington, DC: Government Printing Office.

U.S. Nuclear Regulatory Commission/Federal Emergency Management Agency. 1980. *Criteria for Preparation and Evaluation of Radiological Emergency Response Plans and Preparedness in Support of Nuclear Power Plants.* NUREG-0654. Washington, DC: Government Printing Office.

Valoric, T. S. 1981. "Hazardous Waste Management: Risks, Regulations, Results," *Risk Management* 28:13-18.

Wallace, A.F.C. 1956. *Tornado in Worcester.* Washington, DC: National Academy of Sciences/National Research Council.

Washington Post 1982. "Truths About Nuclear Power," January 28, 1982:A24.

Wasserman, H. and N. Solomon. 1982. *Killing Our Own.* New York: Delacorte.

Weinberg, A. M. 1981. "Reflections on Risk Assessment," *Risk Analysis* I (March): 5-7.

Weinberg, D. 1983. "An Unhealthy Mixture of Science and Politics," *Natural History* 92 (October):12.

Weisskopf, V. F. 1980. "The Double-Edge Sword Called Technology," *Bulletin of the Atomic Scientists* 36 (April):17-21.

Wenner, L. M. and M. W. Wenner. 1978. "Nuclear Policy and Public Participation," *American Behavioral Scientist* 22:277-310.

Westcott, M. R. 1968. *Toward a Contemporary Psychology of Intuition.* New York: Holt, Rinehart, and Winston.

Westing, A. 1978. "Neutron Bombs and the Environment," *Ambio* 7:93-97.

Wetson, G. S. and S. A. Foster. 1983. "Acid Precipitation: What is it Doing in Our Forests?" *Environment* 25 (May):10-12,38-40.

White, G. F. 1945. *Human Adjustment to Floods: A Geographical Approach to the Flood Problem in the United States.* Chicago: Department of Geography, University of Chicago.

White, G. F. (editor). 1974. *Natural Hazards: Local, National, Global.* New York: Oxford University Press.

White, G. F. and J. E. Haas. 1975. *Assessment of Research on Natural Hazards.* Cambridge, MA: M.I.T. Press.

Whittaker, J. et al. 1982. "Risk-Based Zoning for Toxic Gas Pipelines," *Risk Analysis* 2 (September):163-169.

Wiesner, J. B. 1973. "Technology Is For Mankind," *Technology Review* 75 (May):10-13.

Wildavsky, A. 1979. "No Risk Is the Highest Risk of All," *American Scientist* 6 (January/February):32-37.

Winner, L. 1977. *Autonomous Technology.* Cambridge, MA: MIT Press.

Withey, S. B. 1959. "Public Opinion About Science and Scientists." *Public Opinion Quarterly* 23 (Fall):382-388.

Wolpert, J. 1966. "Migration as an Adjustment to Environmental Stress," *Journal of Social Issues* 22:92-102.

Wolpert, J. 1977. "Evacuation from the Nuclear Accident," pp. 125-129 in J. Odland and R. N. Taaffe (editors), *Geographical Horizons.* Dubuque, IA: Kendall/Hunt.

Wolpert, J. 1980. "The Dignity of Risk," *Transactions, Institute of British Geographer* 5:391-401.

Yellin, J. 1976. "The NRC's Reactor Safety Study," *Bell Journal of Economics* 7 (Spring):317-339.
Zeigler, D. J., S. D. Brunn, and J. H. Johnson, Jr. 1981. "Evacuation from a Nuclear Technological Disaster," *Geographical Review* 71 (January):1-16.
Zeigler, D. J. and J. H. Johnson, Jr. 1984. "Evacuation Behavior in Response to Nuclear Power Plant Accidents," *The Professional Geographer* (forthcoming).